GHOST PROTOCOL

GHOST PROTOCOL

DEVELOPMENT AND DISPLACEMENT

IN GLOBAL CHINA

Carlos Rojas and Ralph A. Litzinger, editors

Duke University Press Durham and London 2016

© 2016 Duke University Press
All rights reserved
Printed in the United States of America on acid-free paper ∞
Designed by Amy Ruth Buchanan
Typeset in Minion and Avenir by
Westchester Publishing Services

Library of Congress Cataloging-in-Publication Data
Names: Rojas, Carlos, [date] editor. | Litzinger, Ralph A., [date] editor.
Title: Ghost protocol : development and displacement in global
China / Carlos Rojas and Ralph A. Litzinger, editors.
Description: Durham : Duke University Press, 2016. | Includes
bibliographical references and index.
Identifiers: LCCN 2016004716 (print)
LCCN 2016006130 (ebook)
ISBN 9780822361770 (hardcover : alk. paper)
ISBN 9780822361930 (pbk. : alk. paper)
ISBN 9780822374022 (e-book)
Subjects: LCSH: China—Economic conditions—20th century. |
Economic Development—China. | China—Economic policy.
Classification: LCC HC427.95.G46 2016 (print) | LCC HC427.95
(ebook) | DDC 330.951—dc23 LC record available at
http://lccn.loc.gov/2016004716

Cover art: Lu Hao, *Duplicated Memories* (2008). Installation.
Courtesy of the artist. Photograph by Yomi Braester.

CONTENTS

ACKNOWLEDGMENTS vii

INTRODUCTION "Specters of Marx, Shades of Mao, and the Ghosts of Global Capital"
Carlos Rojas 1

PART I.
URBANIZATION

1. Traces of the Future: Beijing's Politics of Emergence
   Yomi Braester 15

2. The Chinese Eco-City and Suburbanization Planning: Case Studies of Tongzhou, Lingang, and Dujiangyan
   Robin Visser 36

3. Hegel's Portfolio: Real Estate and Consciousness in Contemporary Shanghai
   Alexander Des Forges 62

PART II.
STRUCTURAL RECONFIGURATIONS

4. Dams, Displacement, and the Moral Economy in Southwest China
   Bryan Tilt 87

5. Slaughter Renunciation in Tibetan Pastoral Areas: Buddhism, Neoliberalism, and the Ironies of Alternative Development
   Kabzung and Emily T. Yeh 109

6. "You've Got to Rely on Yourself . . . and the State!":
   A Structural Chasm in the Chinese Political Moral Order
   Biao Xiang 131

7. Queer Reflections and Recursion in Homoerotic
   Bildungsroman
   Rachel Leng 150

PART III.
MIGRATION AND SHIFTING IDENTITIES

8. Temporal-Spatial Migration: Workers in Transnational
   Supply-Chain Factories
   Lisa Rofel 167

9. Regimes of Exclusion and Inclusion: Migrant Labor,
   Education, and Contested Futurities
   Ralph A. Litzinger 191

10. "I Am Great Leap Liu!": Circuits of Labor, Information,
    and Identity in Contemporary China
    Carlos Rojas 205

REFERENCES 225   CONTRIBUTORS 243   INDEX 247

ACKNOWLEDGMENTS

The editors would like to acknowledge the support of the CCK Foundation Inter-University Center for Sinology and Duke's Asia/Pacific Studies Institute for funding a conference at which earlier versions of several papers in this volume were presented. We are also grateful to the staff at the Duke Marine Lab for their hospitality in hosting our conference on their premises. Because the scope of the conference was somewhat broader (and more explicitly transnational) than that of the present volume, we were ultimately not able to include all of the papers, but we are very grateful to Andrea Bachner, Eileen Cheng-yin Chow, Margaret Hu, and Yongming Zhou for their participation and contributions, as well as to Michaeline Crichlow for her support in absentia. In developing the volume itself, we received some exceedingly helpful readers' reports. We are also indebted to Ken Wissoker for his advice and support throughout this process, and to the rest of the staff at Duke University Press. Finally, Ralph Litzinger would like to acknowledge the support he received during his diagnosis of and treatment for cancer during the final stages of this project. In sum, we both appreciate the patience and hard work of all of our contributors, as well as the multiple forms of care—and a good dose of hope—that we never imagined would be required at the start of this project.

INTRODUCTION

"Specters of Marx, Shades of Mao, and the Ghosts of Global Capital"

Carlos Rojas

A specter, one might say, is haunting China.

In *The Communist Manifesto*, Marx and Engels famously use a spectral metaphor to describe the challenge that Europe's mid-nineteenth-century subterranean communist forces posed for the capitalist establishment. Faced with an economic order that was increasingly predicated on the extraction of surplus value from the labor of the proletariat, Marx and Engels called upon their communist brethren to "meet this nursery tale of the specter of communism with a manifesto of the party itself" (Marx and Engels 1998). Here, the "specter of communism" carries a retrospective as well as an anticipatory valence, connoting the past and current suppression of communism and the emancipatory promise that communism holds for the future.

Postsocialist China, meanwhile, may be seen as a mirror image of this vision of mid-nineteenth-century European communism. While *The Communist Manifesto* describes how communism arose in response to the exploitative tendencies of industrial capitalism, in contemporary China it is instead capitalism that is being offered as an antidote to the socioeconomic weaknesses resulting from a quarter century of communist rule. After a Maoist political economic system predicated on egalitarianism yielded widespread poverty punctuated by periods of extreme famine, Mao Zedong's successor, Deng Xiaoping, instead took aggressive measures to encourage rapid economic growth and privatization. Capitalism, in Deng's view, offered contemporary

China an anticipatory promise similar to that which Marx and Engels had previously found in communism.

At the same time, however, this "return" of capitalism to contemporary China is itself haunted by the nation's own enduring socialist legacy. Unlike the Soviet Union and other Eastern European states, whose communist regimes collapsed in rapid succession in the late 1980s and early 1990s, the communist government that Mao Zedong established in China in 1949 was never technically dissolved. Instead, under Deng Xiaoping, China developed a hybrid political economy wherein an embrace of free-market principles was qualified by a retention of many elements inherited from its socialist political structure and social organization. For instance, the country continues to lack open elections, freedom of speech, freedom of the press, or even a fully convertible currency—as a support of capitalism is strategically counterbalanced by a comparatively tight control over many socioeconomic factors (Kirby 2011).

Many socialist-era institutional structures persist in the modern period, though their significance is often greatly transformed. For instance, the household registration, or *hukou*, system was implemented in the 1950s under Mao, and it was initially designed to limit the movement of citizens between different communities and regions in China—and particularly from the countryside into urban areas. Many essential social services were only available to residents of the communities to which they were assigned, and consequently people had very little opportunity to relocate at will. Although the *hukou* system has remained basically intact in the post-Mao era, it is now significantly weakened, and for many rural Chinese the lure of increased earning potential in China's cities outweighs the risk of losing access to the social services accorded them by their *hukou*. As a result, more than a hundred million rural residents are currently living and working away from what were formerly their official home locations. At the same time, however, while the *hukou* system no longer prevents internal migration, its vestigial presence does continue to carry significant social consequences, because it relegates China's migrant workers to the status of partially disenfranchised second-class citizens (Young 2013).

This interweaving of earlier socialist structures with newer capitalist ones may be seen as an institutional version of a phenomenon that the historian Geremie Barmé (1996) has memorably called "shades of Mao." Barmé describes how, in the 1980s and 1990s, there was a veritable explosion of Maoist images and icons in China, as Mao Zedong—who, during his reign as the titular head of the People's Republic, had encouraged a cult of personality centered around himself while at the same time attempting to abolish both organized and folk religious practice—ironically became transformed into a virtual deity in his own right, and Maoist icons were treated as sacred relics. The contemporary fascination with these icons, therefore, marks an uncanny return of the Great Helmsman's legendary cult of personality, while also embodying the same religious iconography and commodity fetishism that Mao had originally tried to abolish. By a similar token, the market capitalism that emerged in China following Mao's death is closely intertwined with his legacy, as the social institutions he helped put in place have become potent symbols of both China's socialist past as well as its capitalist present.

Finally, the partial displacement of communism by the forces of global capital has helped bring into sharper relief the Marxian critique of capitalism that had inspired China's communist system in the first place. That is to say, China's rapid economic growth in the post-Mao period has not only brought hundreds of millions of Chinese out of poverty, but also generated yawning disparities of wealth and earning potential. The resulting witch's brew of state-owned enterprises, joint-venture companies, wealthy investors, middle-class consumers, and a vast floating population of migrant labor has yielded a highly fluid relationship between capital and labor that virtually demands a critical Marxian analysis.

In his influential study *Specters of Marx*, Jacques Derrida argues that, in the wake of the collapse of most of the world's communist states, the *spirit* of Marx's original critique remains more relevant than ever. He contends that the spread of neoliberalism has contributed to an epidemic of "violence, inequality, exclusion, famine, and thus economic oppression" around the world, and suggests that Marxian theory offers a productive response to these problems (Derrida 1994: 106). In particular, Derrida contrasts the way foreign debt is mobilized to support the legitimacy of a neoliberal regime, on one hand, and the way a logic of intellectual *indebtedness* to the critical tradition associated with Marxism necessarily underlies any contemporary attempt to reappropriate a legacy of Marxian critique, on the other.

In this volume, we argue that in contemporary China the structural tensions produced by overlapping socialist and capitalist regimes have the potential to open up a set of new emancipatory possibilities. In short, we argue that present-day China is shadowed by three mutually imbricated sets of spectral apparitions: (1) a capitalist order that promises to reverse the consequences of China's quarter century of Maoist rule; (2) the institutional legacy of this Maoist regime, which continues to shape contemporary Chinese society in critical ways; and (3) a spirit of resistance and critique consonant with the underlying principles of Marxist theory. Contemporary China, in other words, is haunted by what we might call the ghosts of capital, shades of Mao, and specters of Marx.

Some of the implications of this tripartite haunting become more apparent if we return to Derrida's *Specters of Marx*. Despite having been written at a time when the People's Republic of China was one of the world's few remaining communist states, Derrida's study makes virtually no mention of China, though the work does carry traces of China's latent presence in the form of three allusions embedded in the margins of the text, each of which is associated with a different valence of contemporary China's post-Maoist condition.

To begin with, China is mentioned only once in *Specters of Marx*, and this reference is not even in Derrida's own voice. Instead, it appears in Derrida's discussion of the famous analysis of commodity fetishism in *Capital*, in which Marx posits that material commodities come to assume a mystical character that occludes the human labor and social relations responsible for their production. Marx cites as an example a wooden table, noting that as a commodity it comes to assume anthropomorphic qualities "far more wonderful than 'table-turning' ever was," and in a corresponding footnote he observes that "one may recall that China and the tables began to dance when the rest of the world appeared to be standing still" (Marx 1977: 163; cited in Derrida 1994: 153). Here, Marx is evidently alluding both to the fascination with spiritualism (and specifically "table-turning") that developed among the European aristocracy following the failure of the continent's revolutionary movements of 1848 and 1849, as well as to China's Taiping Rebellion (1850–1864), which nearly toppled the Qing dynasty during the same midcentury period. Given that the Taiping Rebellion embraced such progressive goals as gender equality, the abolition of private property, and the equal distribution of land, the rebellion may be seen as a reflection of the objectives outlined in *The Communist Manifesto* itself, and therefore Marx's image of midcentury China "dancing" while "the rest of the world appeared to be standing still" positions the Central Kingdom as a spectral obverse of European society—or what Marx ironically calls here "the rest of the world."

As with China, the word *Chinese* is also mentioned only once in Derrida's text, and this reference is not in Derrida's own voice either. Rather, it appears in his citation of Alexander Kojève's postscript to the second edition of his *Introduction to the Reading of Hegel: Lectures on the Phenomenology of the Spirit*, in which Kojève proposes that "if Americans appear to be Sino-Soviets who have gotten rich, it is because the Russians and the Chinese are still but impoverished Americans, moreover on a rapid road to wealth" (Kojève 1962: 436–437; cited in Derrida 1994: 72). Writing in 1962, Kojève, in this eerily prescient passage, sees the Soviet Union's and Maoist China's emphases on poverty and the redistribution of wealth as an uncanny anticipation of the "rapid road to wealth" on which he believes both nations will soon embark. He suggests, in other words, that communism contains buried within itself a capitalist imperative against which it is nominally opposed.

A third allusion to China may be found not in Derrida's text proper, but rather in the work's historical margins. When Derrida first published *Specters of Marx* in 1993, many of Asia's economies had been enjoying several years of impressively robust economic growth, though only four years later a chain reaction of currency collapses in several Asian countries temporarily brought this growth to a virtual standstill. While the precise causes of this regional crisis are complex and multifold, one important factor was a collective anxiety surrounding the impending return of Hong Kong to Chinese control. The Hong Kong Retrocession was perceived as having crucial implications not only for Hong Kong but also for the region as a whole, insofar as much of the region's foreign investment was channeled through Hong Kong financial institutions. It is telling, therefore, that the first widely recognized symptom of the Asian financial crisis—the abrupt devaluation of the Thai baht as a result of aggressive selling on the international currency markets—occurred on July 2, 1997, just one day after the July 1 return of Hong Kong to China. In the following weeks, several other Asian countries similarly suffered runs on their currencies, as the entire region quickly found itself mired in an acute recession. China, however, remained largely unaffected—in part because its currency was not fully convertible and, hence, was shielded from the international currency market. To the extent that the Asian miracle may be seen as an expression of a capitalist logic, China's role in helping to provide the catalyst for the resulting crisis—together with the nation's own subsequent immunity from the most serious consequences of the crisis itself—illustrates the role that the nation's vestigial communist institutions and policies continue to play in the nominally postsocialist world.

One of the international community's primary responses to the Asian financial crisis, meanwhile, was to arrange for the International Monetary Fund (IMF) to grant large bailout packages to help stabilize the currencies of the countries most severely affected by the crisis. The IMF, following a practice initially developed in the 1950s, made the loans contingent on the countries' agreements to reform their macroeconomic and monetary policies in accordance with IMF demands. This practice of "conditionality," however, has been criticized for being insufficiently attuned to the specific needs of each individual country or region. The IMF has been accused, in other words, of serving as a vehicle for the export of a tacitly imperialistic model of Western capitalism, and the Asian financial crisis dramatizes the tension between the global spread of capitalism and economic growth, on one hand, and Western-centered financial institutions' attempts to circumscribe this proliferation of capitalist models, on the other.

Just as Derrida's seminal study *Of Grammatology* characterizes the Chinese script as "a sort of European hallucination" for Enlightenment-era thinkers attempting to critique the perceived limits of European writing systems (Derrida 1998: 80), in *Specters of Marx* China itself functions as a hallucinatory mirror image of Western capitalism. In each of the three preceding allusions ensconced in the margins of Derrida's text, China—either in its precommunist, high communist, or "postcommunist" incarnations—is viewed as not only an antithesis of capitalism but also an ironic embodiment of capitalism's own underlying logic. In particular, and to use the conceptual shorthand introduced above, Marx's "dancing China" allusion to the Taiping Rebellion articulates a desire for an alternate expression of Marx's own communist ideals—which is to say, the rebellion functions as an ironic anticipation of what Derrida calls the specters of Marx. Kojève's "road to wealth" discussion of the People's Republic, meanwhile, describes how even Maoist China was haunted by a suppressed desire for wealth, or what we might call the ghosts of capital. Finally, contemporary China's partial complicity in—and its subsequent immunity from—the 1997 Asian financial crisis may be seen as an illustration of the institutional legacy of communism, or what we are calling here the shades of Mao.

To the extent that Derrida's *Specters of Marx* argues that the contemporary neoliberal order has produced a set of socioeconomic relations that invite a Marxian critique, the preceding reading of the virtual absence of China from Derrida's own text suggests instead that the nation's peculiar historical trajectory in the post-Mao era may offer an alternate entry point into an assessment of the relationship between capitalism and communism. Here, accordingly, we are interested in the way contemporary China illustrates some of the concerns that haunt the margins of Derrida's text.

In this volume, we take as our starting point China's rapid transformation from an impoverished communist state into the world's largest economy, and we approach economic development as the result not of a set of ineluctable historical imperatives (as imagined within Marxism) or deliberate economic strategies (as imagined within capitalism), but rather as discrete practices and procedures, or what we call here "protocols." The product of overlapping institutional strategies, political procedures, legal regulations, religious rituals, and everyday practices, these protocols are significant for their real-world implications as well as for the ways they may help shape a public imaginary. They not only have a potentially normalizing function, helping interpellate subjects into a national political economy, but they may also encourage a pattern of disruptive, counterhegemonic tendencies that may in turn help generate new subject positions that potentially stand at cross-purposes with the nominal objectives of the national political economy. We are particularly interested in the way these protocols may help constitute national subjects and the nation itself, while at the same time they may also create a space wherein this same national imaginary may be interrogated. Drawing from methodologies associated with both the humanities and the interpretive social sciences, we seek to situate contemporary China's dialectic of development and displacement at the juncture of society and ideology—between the world as experienced and the world as imagined.

When Deng Xiaoping emerged as China's de facto leader following Mao Zedong's death in 1976, he found himself at the helm of a nation that was predominantly poor and rural, and was comparatively isolated from the West. Deng's Reform and Opening Up campaign in 1978 was designed to help jump-start the country's economy, and it succeeded in a spectacular fashion. Over the following three and a half decades, China's GDP increased more than a hundredfold, and in 2014 the nation passed the United States to become the world's largest economy. During the same period, China's foreign exchange reserves increased by a factor of more than two thousand, from $1.6 billion to more than $3.2 trillion, giving it the largest store of foreign exchange reserves in the world (and more than its four closest rivals—Japan, the European Union, Saudi Arabia, and Russia—combined).[1] The result is a peculiar hybrid—a capitalist juggernaut retaining elements of a socialist planned economy—that has proven particularly adept at weathering the global financial crises that have rocked many of the world's other economies.

Our goal, however, is not to celebrate contemporary China's rapid growth, but rather to critically examine its structural contradictions. While China's economy has grown rapidly since the early 1980s, this new wealth has been very unevenly distributed. Indeed, at the beginning of the Reform Era, Deng Xiaoping famously declared that a minority of the population should be allowed to "get rich first," and sure enough the nation's Gini coefficient—a standard measure of income inequality—has subsequently ballooned from an impressively low 0.16 to a dangerously high 0.47. The result is a nation marked by a set of profound contradictions. China has one of the world's largest and fastest-growing markets for high-end luxury products, but it also has one of the world's largest pools of cheap labor. It has many of the world's largest and fastest-growing metropolises, yet as many as a third of the residents of its largest cities are migrant laborers with little legal standing. It has one of the world's largest and fastest-growing telecommunications industries, yet it is notorious for its attempts to exercise political control over public and private expression. Contemporary China, in short, represents a hyperbolic expression of both global capitalism *and* its putative communist antithesis, and it is within the interstices between these two nominally antithetical political-economic regimes that one finds an array of new structural possibilities.

To explore these tensions, we offer a series of case studies examining the relationship between contemporary China's economic development and its ramifications—particularly the relationship between capital and labor, governmentality and biopolitics, regionalism and globalism. Through the lens of various social and cultural phenomena, we consider some of the ways China's growth has been perceived and experienced by those most immediately affected by it, arguing that an attention to these *perceptual realities* offers the possibility of a more nuanced understanding of the political and economic protocols that help shape this growth.

This volume is divided into three parts, focusing on interrelated themes of urbanization, transformation, and migration. In part I, we begin by examining the figure of the city, as both a material place and an imaginary entity. According to official census figures, between 1982 and 2010 the percentage of China's population living in urban areas expanded from 20.6 percent to 49.7 percent. Given that the nation's overall population grew by more than a third of a billion during the same period, the total increase in urban residents was more than 450 million people. The result has been a rapid increase in the size and number of the nation's urban centers, to the point that China currently has more than 120 cities with populations of greater than a million, including several in excess of 10 million. One consequence of China's rapid urbanization, however, is that many of its cities now find themselves wedged between memories of their past and anticipatory visions of the metropolises that they are on the verge of becoming. The significance of the city, accordingly, lies in its status as not merely a material entity and a lived space, but also as a symbolic space conditioned both by a collective recollection of the past and an anticipation of the future.

The three chapters in part I—by Yomi Braester, Robin Visser, and Alexander Des Forges—examine processes of urbanization in contemporary China. Focusing on themes of urban construction, the construction of "new cities" at the geographic peripheries of existing metropolises, and the relationship between cities and their residents, these three authors reflect on some of the implications of the rapid growth and transformation of urban regions in contemporary China. They argue that cities are frequently viewed through the lens of a future that they have not yet become and that this future-oriented perspective has complex—and often quite problematic—implications for the politics of urbanization discourses and their corresponding practices.

Next, we turn our attention to the ways in which ordinary citizens position themselves within a contemporary landscape characterized by competing socialist and neoliberal orders. Under Mao, the state was often quite up front in its attempts to shape the opinions and beliefs of its citizens, and while these efforts have continued in the post-Mao era they often overlap with sociopolitical, cultural, and religious forces that have the potential to interpellate subjects in different ways.

The four chapters in part II—by Bryan Tilt, Kabzung and Emily T. Yeh, Biao Xiang, and Rachel Leng—examine different communities and demographic groups positioned at the outer borders of China's contemporary social order. From ethnic minorities in Southwest China's Sichuan Province to yak-herding communities in Tibet, and from economically depressed communities in China's far northeast to queer subjects positioned at the symbolic margins of its major cities, these four case studies focus on groups that find themselves at the interstices of these conflicting developmental imperatives. Each of these analyses, however, contends that these overlapping configurations can potentially generate new possibilities for individual self-positioning and self-constitution, as these subjects may leverage different sociopolitical forces to create new positionalities.

Finally, we turn to China's sea of migrant workers. The growing economic disparities between urban and rural areas have produced a set of powerful incentives drawing workers from the countryside to the cities. Overwhelming the Maoist-era *hukou* system, this wealth gap has generated a vast floating population of rural-to-urban migrants. China's 2010 census found that more than 220 million people had spent at least six months of the preceding year living away from their hometowns, including approximately 130 to 160 million rural-to-urban migrant workers. In fact, this floating population of migrant workers is currently so large that if it were a nation unto itself, it would be the tenth most populous country in the world. It is, however, a population that lacks many of the rights and privileges enjoyed by official urban residents, and the result is a situation in which migrant workers provide much of the cheap labor on which the nation's continued economic growth is predicated, even as many of them are condemned to a precarious existence and enjoy only limited benefits from the growth they helped make possible in the first place (Lee 1998, 2007).

Rather than treat this floating population of migrant workers as a strictly demographic phenomenon, however, we focus instead on how these workers are classified by the socioeconomic systems within which they are positioned and how they themselves appear to understand their own position within these same systems. Looking at workers in Italian-owned garment factories, at the school-aged children of migrant workers, and at fictional migrants caught up in complicated confidence schemes, the final three chapters in this volume—by Lisa Rofel, Ralph Litzinger, and Carlos Rojas—examine different populations of migrants whose subjectivities are shaped by the overlapping sociopolitical structures within which they are embedded. In particular, each of these groups is driven by a set of aspirational dreams of what they hope to become, though often this optimism may well have a counterproductive—or, as Lauren Berlant (2011) calls it, a "cruel"—dimension, in that their dreams of the future may in fact directly undermine their ability to thrive in the present.

Unlike socioeconomic processes that are perceived as being rooted in either Marxist historical determinism or capitalist imperatives, the protocols that we examine in this volume are often highly contingent and provisional. To the extent that these protocols derive out of the institutional and ideological fractures that open up between the ostensibly antithetical regimes of Maoism and neoliberalism, we contend that they are inflected both by a logic of global capitalism and by a set of institutional structures developed during the socialist period, while at the same time they have generated a set of conditions that invite a neo-Marxian critique. We are interested, in other words, in a body of protocols, haunted by the ghosts of capital, shades of Mao, and specters of Marx, that collectively suture China's future to its past and present and also help to negotiate the nation's relationship with the rest of the world.

In the 2011 blockbuster film *Mission Impossible—Ghost Protocol* (directed by Brad Bird), the members of a special ops team are unexpectedly implicated in a mysterious terrorist bombing of the Kremlin. In response, the U.S. president activates a top secret policy referred to as "Ghost Protocol," whereby the team is officially disbanded while simultaneously instructed to continue working in an unofficial capacity to track down those responsible for the bombing. The result is a delicate dance of deception and duplicity, in which the team is stripped of its legitimacy, authority, and even its very identity so that it may attempt to locate the covert enemy that had framed it.

Although Bird's film is obviously fantasy, it nevertheless resonates with contemporary concerns in suggestive ways. For instance, the covert team is called IMF, which is short here for Impossible Missions Force but is, of course, also the acronym for the International Monetary Fund whose function, like Hunt's fictional team, is to diffuse existing crises and prevent new ones. Also, Bird's film revolves around a set of Cold War themes that include covert Russian agents, stolen missiles, and the threat of nuclear war. By resurrecting the former Soviet Union to serve as a worthy antagonist, the film tacitly acknowledges the continued relevance of the communist ideology that the USSR represented. Furthermore, the two years preceding the release of *Ghost Protocol* coincided with emerging concerns that Greece, one of Europe's most rapidly developing economies in the early 2000s, was having trouble financing its growing debt. Despite repeated bailout attempts, Greece's economy has continued to spiral out of control, precipitating a financial crisis that threatens the very existence of the European Union itself. Hidden in the margins of Bird's film, accordingly, we find allusions to the International Monetary Fund, the former Soviet Union, and contemporary Europe's sovereign debt crisis, suggesting the degree to which the contemporary world remains haunted by the ghosts of capital, the shades of communism, and the specters of Marx.

The circumstances surrounding Europe's contemporary debt crisis mirror those of the 1997 Asian financial crisis, in that, in both cases, a combination of debt and inflation threatened to destabilize an entire region's economy. If the Asian financial crisis was largely rooted in the fact that the currencies of the various countries were too *independent* of one another (thereby permitting speculators to play one currency against the other), however, the European sovereign debt was instead a product of the continent's decision to unify behind a single currency. While the IMF's intervention in the Asian financial crisis was criticized as a form of financial imperialism, in the European instance we find an inverse situation, in that in June 2012 China announced that it was leading the so-called BRICS nations (Brazil, Russia, India, China, and South Africa) and about fifteen other developing nations in making new funding pledges to help the IMF address Europe's debt crisis. In return the countries demanded greater representation and influence within the fund itself. In this way, China was attempting to capitalize on Europe's economic crisis in order to leverage a greater voice for itself within the IMF, thereby figuratively colonizing the European-centered global monetary institution from within.

This mutual invagination of capitalism and its spectral Other, meanwhile, captures the analytical logic of our volume, which focuses on a set of ghostly protocols haunted by intersecting and conflicting ideological and programmatic imperatives. Just as the specific contours of post-Mao China's embrace of capitalism continue to be shaped by the country's vestigial socialist institutions and ideologies, those same socialist institutions are *themselves* inflected by the shadowy presence of capitalism and capital (either in the form of a historical legacy, à la Marx, that communism was originally attempting to displace, or as an optimistic anticipation, à la Kojève, of the hyperbolic consumerism against which communism is theoretically opposed).

Note

1 See the CIA's "World Factbook: Reserves of Foreign Exchange and Gold," and the IMF's "Data Template on International Reserves and Foreign Currency Liquidity—Reporting Countries."

PART I

URBANIZATION

TRACES OF THE FUTURE

Beijing's Politics of Emergence

Yomi Braester

Lu Hao's installation *Duplicated Memories* (2008) presents two versions of Beijing. One, on the ground, is a large (about eighty square meters), backlit, contemporary transportation map of Beijing's Old City (see image on the cover of this volume). Another, hovering in the air, consists of Plexiglas maquettes of the Old City gates, most of which were demolished by the 1970s. Lu shows the past not as buried under the city, but rather as hovering above, a ghostly reminder that replicates, complements, and challenges the everyday practices of moving in the city. Lu, who had previously replicated the Tiananmen Gate as a Plexiglas fishbowl (*Fish Bowl/Tiananmen*, 1999), imagines the Old City gates not in concrete architectural terms but rather as shimmering artifacts that require the viewer to look up and away from the city map and renegotiate one's place in relation to the city and its double. When the installation moved to the OCT Contemporary Art Terminal in Shenzhen in 2009, it was accompanied by video art projected around it, further duplicating the urban environment within additional media.

Lu's artwork is symptomatic of the discourse on Beijing's modernization since the late 1990s. The capital has undergone massive demolition, construction, and gentrification, and its skyline has been redefined by brand-name landmarks. The visitor to Lu's installation is asked to imagine Beijing's past, present, and future as defying the onslaught of material destruction and converging, in a remedial gesture, in collective memory. As in other quickly developing cities, a prevalent trope for representing Beijing and imagining its redemption is that of the palimpsest. Like erased ink surfacing onto scraped sheepskin of the palimpsest to show traces of earlier writings, goes the claim, so do earlier urban patterns affect citizens' perceptions of the present. The palimpsest is a tool for the spatialization of memory—that is, it places the passing of time within a narrative intelligible through reference to the transformation of space. By espousing the idea of the palimpsest, urban dwellers can easily anchor their identity in time and space, resulting in their empowerment as self-conscious subjects.

Although I acknowledge the transformative potential of the palimpsest and related metaphors, I propose that they can also manipulate collective memory by allowing the citizen to be subsumed by hegemonic formulations of subjectivity. The figure of the palimpsest seems on its face to offer a location-specific and historically coherent self, but in stressing the layering of time and space the metaphor is often employed to justify the fragmentation of experience and temporal disorientation, in effect introducing collective amnesia. This cover-up is motivated by capitalist interests and has immediate on-the-ground implications. The palimpsest adds a veneer of historical continuity that expedites the commodification of sites and their integration into neoliberal economy.

This chapter explores the ideological and material apparatuses of visualizing the historically layered city, as manifest in present-day Beijing. I start by historicizing the concept of spatial and temporal overlaying and eliciting its critical potential and fault lines. I use for a case study the reconstruction and gentrification of Beijing's Qianmen district in 2008 to show how street spectacles have redefined the temporal significance of the built environment. Buildings, billboards, and digital screens have formed contiguous media in the service of urban utopia. Mourning the disappearing cityscape, and the so-called culture of disappearance, is complemented by what I call the *politics of emergence*—celebrating new construction and projecting an anticipated future onto the perceived present. Urban discourse in the service of real estate developers fashions the city as a palimpsest bearing traces not only of the past but also, more importantly, of the future.

The implications for fashioning Beijing as a palimpsest bearing traces of the future are tangible and material. I have argued elsewhere that visual practices have accompanied urban development in China and shaped it. Texts, images, maps, and buildings coalesce in creating chronotopes—imaginary space-and-time frames for interpreting places and their historical significance (Braester 2010a). Screen cultures have facilitated the redaction of city plans, the redefinition of public spaces, and the reassessment of cultural heritage sites and political monuments.

Since 1949, Beijing has been imagined through various chronotopes that have fashioned its future as lying on its doorstep. The national and municipal governments have provided idealized images that prescribed the normative uses of space. To emphasize the urgent need for urban reform and the authorities' adequate response, plays and films have portrayed the change as taking place overnight, thereby skirting the painful process of demolition-and-relocation. The prescriptive chronotope and the chronotope of instantaneity, as I have called them, have been transformed more recently into the postspatial chronotope, which abolishes material space altogether by bringing social interaction to virtual environments and augmented reality. The temporal merging and telescoping implicit in these practices are now made more explicit. The conceptual matrix for Beijing's space-time in the twenty-first century, under the aegis of the politics of emergence, fashions the present as a placeholder for things to come. The protocols of engaging with public space are seen as the ghostlike effigy of what is yet to emerge. Current spatial practices can be understood only insofar as they are palimpsestic traces of a utopian future.

The discourse that identifies the urban texture as a palimpsest mummifies the past and fetishizes the future. The nostalgia for historical splendor, and its mirror image in yearning for grandeur to come, covers up for a troubled present. Claiming that the present is knowable only as the future's past obfuscates the immediate functions of capital. The spatial and temporal fragmentation of the present—the violent denial of its representability—is the by-product if not the intended result of subjecting the built environment to market economics. Developers lionize the city as a historical spectacle in order to fashion space as a site of unachieved promise. The logic of concealed potential and intrinsic incoherence belittles the significance of the readily visible urbanism as shaped by neoliberal policies. The politics of emergence diverts attention from the city at hand as the spatial articulation of unchecked capitalism. Populating the present with specters of past and future temporalities serves the dominant economic and political powers.

In pointing out the inviability of utopian claims in present-day Beijing, I do not dismiss them as simply naïve or insincere. The unquestionable propagandistic use of spatial and media practices notwithstanding, these practices are symptomatic of the failure of the historical imagination under the current social and economic conditions. David Harvey (2000b) has described the New Urbanism as "degenerate utopias." The New Urbanism creates gated and surveilled theme park–like, self-contained bubbles, based on a quasi-mythical historical view of the city. Modernist idealism has been reduced to utopia by architectural design, dedicated to facilitating commodity culture.

Palimpsest and Simulacrum

The idea of the palimpsest, as applied to the modern city, has defined urban thought throughout the twentieth century. A short history of the concept may explain its attraction for urbanists as well as the dangers in its uncritical use. The notion that the city exists simultaneously at multiple symbolic and material levels may be traced back to Babylonian cosmology and the rabbinic vision of Celestial and Earthly Jerusalems.[1] In the nineteenth century, the palimpsest became a common metaphor for human consciousness and for history (McDonagh 1987). Sigmund Freud first connected the palimpsest to the urban layout, comparing Rome and its layered history with the human brain, "in which nothing that has once come into existence will have passed away and all the earlier phases of development continue to exist alongside the latest one" (1962: 17; see also Freud 1959). Since Freud, the urban palimpsest has carried an implicit analogy to modern subjectivity.

Michel de Certeau brought the term to wide use when he denoted a place laden with memories as a palimpsest (1984: 109). In his famous "Walking in the City" as well as in the seminal essay "Ghosts in the City," Certeau fashions the built environment as a cipher made of historical layers. Following Walter Benjamin, Certeau focuses on the flesh-and-blood individual walking the city among a noisy, sweaty, and heaving crowd, and he privileges that individual over the abstract constructs of "the geometrical space of urbanists and architects" (1984: 100). The two spatial forms interpenetrate, and the modern citizen's task is to make sense of their simultaneous existence. The palimpsest-like nature of sites renders them, in Certeau's words, into "fragmentary and inward-turning histories, pasts that others are not allowed to read, accumulated times that can be unfolded but like stories held in reserve, remaining in an enigmatic state" (1984: 108; see also de Certeau 1998). Certeau opposes planners' attempts to begin with a clean slate that avoids a necessary ambivalence in understanding space. Instead, he presents walking in the city as an act of textual reading that excavates hidden narratives. The people walking the streets become citizens and gain their subjectivity from telling the city's stories. Walking in the city is a Heideggerian act of constituting one's being-there. The ability to extricate meaning out of space and make the city livable depends on citizens' self-conscious acts of invoking the past.

Certeau's idea of the palimpsest might be seen simply as rejecting idealized spaces and as emancipatory of ideological totality: It benefits the subject who teases meaning out of the unsaid and unbuilt, divulges the deeper structures of the city, and frees the citizen to create a cognitive map (Buchanan 2000: 121–122; Jameson 1991: xi, 51–52). Even David Harvey, who identifies the palimpsest with a capitalist construction of "social forms . . . in the images of reciprocity, redistribution and market exchange," seems to ignore his own reservations and describes the urbanist's task as that of preserving the palimpsest: "Urban designers . . . face one common problem: how to plan the construction of the next layers in the urban palimpsest in ways that match future wants and needs without doing too much violence to all that has gone before" (2009: 245; Harvey 2000a: 27–28). Probably the most sustained interest in the emancipatory potential of the palimpsest is found in Andreas Huyssen's *Present Pasts: Urban Palimpsests and the Politics of Memory*. Huyssen cautions that the recent "memory fever" may facilitate collective amnesia. For Huyssen, the palimpsest is a critical tool that responds to unreflective nostalgia (2003: 1). The palimpsest, in his view, enables spatial inquiry and at the same time privileges visual media as the means of such investigation.

And yet the concept of the urban palimpsest can both counter and support hegemonic discourse. Certeau, amid his enthusiastic endorsement of the urban palimpsest as an ideological tool, warns against using the palimpsest for a selective, fetishizing representation of the city as a showcase of the past. Fashioning the urban environment as a palimpsest might turn the city into a museum, one that "conceals from users what it presents to observers . . . pull[ing] objects away from their everyday use" (1998: 138; see also Buchanan 2000: 19–21). The excavation and uncovering must be taken up by the citizen, not presented in preprocessed form, ready for consumption. Inquiry into the palimpsest must avoid turning the city into an idolized object, an immutable exhibit.

Christine Boyer, inspired by Pierre Nora, adds a warning—perhaps too dystopic—about the palimpsest's role in distorting memory.[2] Rather than enhancing collective memory, the concept of the palimpsest ends up creating a disconnect between past and present. Insofar as monuments, restored sites, and artworks purport to function as "environments of memory," they do so only to satisfy a touristy gaze, one that does not take root in everyday activities but rather caters to the commodification of sites (see Boyer 1994: 339). The palimpsest is used to refer uncritically and ahistorically to invented traditions.

Boyer follows the bathos of Jean Baudrillard's criticism of contemporary urbanism, in which the latter identifies the "end of representation and implosion ... of the whole space in an infinitesimal memory, which forgets nothing, and which belongs to no one" (Baudrillard 1994: 71). For Baudrillard, the city is not the sum total of its pasts and presents. It does not retain memory or contain the figure of redemption. One may add: In its failure to produce a functional space, the urban simulacrum is abetted by the logic of the palimpsest, which may easily recede into infinite replication of the site, an aporia that voids referentiality and defers indeterminately spatial and temporal specificity.

The Baudrillardian critique of the mimetic order in urbanism has brought forth alternatives to the palimpsestic city. In tribute to Baudrillard, Edward Soja defines the contemporary city—the "postmetropolis"—as a place guided by the nonrepresentational logic of the simulacrum. In this "simcity," "simulations of a presumably real world increasingly capture and activate our urban imaginary and infiltrate everyday urban life" (1997: 27). Also taking his cue from Baudrillard, Marc Augé has historicized the transformation of spatial semiotics in the "supermodern" age. Augé designates sites that resist spatial and historical specificity as "non-places." Whereas place "is never completely erased," the contemporary, devalued non-places are "never totally completed; they are like palimpsests on which the scrambled game of identity and relations is ceaselessly rewritten" (Augé 1995: 79). Augé leaves his remark at that, but it carries important implications: The palimpsest is oriented toward the future as much as toward the past; it is part of a dynamic discourse, which must generate time and again imaginary spatial matrices to avoid recognizing the spatial uniformity created by globalization. Insofar as the palimpsest is still useful for conceptualizing the city, it is by drawing attention to how the city is restructured to skirt the burden of the past and refer to the future alone.

Beijing: Palimpsest City

Despite growing reservations among key thinkers about the city's function as a system of signification through superimposed temporal and spatial references, Chinese urbanists have emphatically fashioned cities as exemplars of semiotic layering. China's major cities have been branded, by developers and countercultures alike, through invoking the logic of the palimpsest. As Beijing undergoes rapid demolition and construction, the prevalent discourse has stressed the capital's changing façades. Popular and scholarly books about the city's architectural history, anthologies of old photos of buildings long gone, street displays of former layouts, stage plays showing semi-extinct ways of life, fiction films and documentaries recording the disappearing cityscape—all these have created a cartography that juxtaposes Beijing's past and present (Braester 2010a). The term *palimpsest* (*chongxieben*) is rarely invoked explicitly, and when it is, it is mentioned mostly by urbanists—usually foreign architects who have set up shop in Beijing.³ Yet the implications of the urban palimpsest—viewing the built environment as a set of layers awaiting excavation—have dominated public discussions and urban practices.

The most prominent critic to address the temporal coding of Chinese urbanism is Ackbar Abbas, who has advanced a postmodern semiotics informed by Jacques Lacan, Jean Baudrillard, and Paul Virilio. In *Hong Kong: Culture and the Politics of Disappearance*, Abbas observes Hong Kong culture gearing up for the 1997 handover to the PRC. Hong Kong as people knew it was about to disappear, and they were looking to quickly invent a past for themselves. In following essays, Abbas explores the spatial and temporal signification in other Chinese cities, including Beijing (Abbas 1997, 2003, 2008). For Abbas, the Chinese city at the end of the twentieth century is a sign without any fixed signifier, defined by its inherent unpredictability, an urbanism that is essentially unknowable, paradoxically rooted in the not-here and the not-now. The urban environment constitutes, so to speak, an aporia of the sublime: one unrepresentability couched inside another, a temporal unknown represented through another. Abbas is interested in the multiplication of referents and often invokes the term *overlaying*. But for him the superimposed elements do not reaffirm each other. Instead, Abbas focuses on the reference-less, the patently fake, and the façade, and he regards the new Asian city as an exemplary "any-place-whatever" (Abbas 2008: 243–244).

In tribute to Abbas, I suggest that the culture of disappearance is mirrored by a *politics of emergence*. Real estate developers, with the authorities, celebrate new construction, emphasizing the immediate gratification of an instant city. They employ a rhetoric of urgency that promotes rapid development. Rem Koolhaas, who designed the CCTV Tower, one of Beijing's celebrated landmarks, has famously championed around-the-clock, no-red-tape construction in the PRC, calling for architecture as a "fast medium" (Mattern 2008). The recent real estate bubble in phantom cities like Ordos provides another example of the politics of emergence, which projects rapid development in advance of a city's population if not construction. Hyperurbanization is celebrated as if the emergence of a new city does not come at a cost—often that of the old one's disappearance. The palimpsest ensures the museumified endurance of the past and promises continuity into the future. A brave new world glimmers through semitransparent vestiges of other temporalities.

The politics of emergence is especially manifest in public displays that anticipate the transformation of specific sites. The cityscape has been splashed with billboards and screens that announce construction and simulate its future outcome. Large-scale building projects, often involving the demolition and gentrification of entire districts, have been implemented since the late 1980s. Real estate developers soon saw the advantage of the walled perimeters of these construction sites as advertisement space, often promoting their own projects.

The ideological implications of these promotion practices, and their manifestation in urban structure and visual media, are illustrated by the growing symbiosis between buildings and screens in contemporary Beijing. Even by today's standards of ubiquitous visual stimulation, walking in Beijing provides a bewildering succession of encounters with still and moving images. As advertisers look for maximum visibility and developers aim for spectacular effects, they flaunt ever larger and more technologically advanced imaging. The Place, a shopping mall in Beijing's Central Business District (opened in 2007), takes pride in a 220-by-27-meter skyscreen, the second-largest in the world. (Like the world's largest screen, at Las Vegas's Fremont Street Experience, the screen at the Place was designed by Jeremy Railton.) The Xicui Entertainment complex (opened in 2008) sports a 2,200-square-meter zero-energy media wall (designed by Simone Giostra & Partners), the largest color LED display worldwide and the first photovoltaic system integrated into a glass curtain wall in China. For the 2008 Olympics, Zhang Yimou utilized a 147-by-22-meter flexible LED screen. The Supertron screen set up from time to time at Tiananmen Square is smaller in comparison but deserves mention for its prominent and symbolic placement, at the center of the capital.

The phenomenon is not confined to Beijing, of course. It may be traced to the introduction of the jumbotron in the mid-1980s and large LED screens in the late 1990s, which accelerated the symbiosis between cityscapes and screen cultures. Times Square in New York and Shibuya in Tokyo have pioneered an ecosystem that does not differentiate between built environments and televisual spectacle. Likewise, the pedestrian street at the center of Guangzhou sports multiple screens, and the Wujiaochang Wanda Plaza in Shanghai (opened in 2006) is designed to double as abstract video art. These screens not only inundate urban space with moving images, they also designate public space as such through the presence of large screens, and define the citizen as a viewer of such screens. The screen walls facilitate a specific relation between urban space and time, whereby the city is a sign of its own future.

Paul Virilio noted that "in some Asian cities . . . façades [are] entirely made of screens. In a certain sense, the screen became the last wall."[4] The symbiosis of architectural and imaging devices, to the point of their conflation, is more than the result of converging technologies. When screens become buildings and buildings become screens, the viewer is made to relate to the images by walking along them and among them. The visual environment of these screens creates for the citizen a time bubble, sequestered from other temporalities and defined solely by the act of viewing. The convergence of architecture and screen is a symptom of how urban spaces can fold unto themselves, doubling as their potential other avatars, on other temporal planes.

The confluence of wall and screen underscores the importance of screen media—from film and video to advertisement boards and smartphones—in shaping public space as layered. As James Tweedie and I have argued, the metropolis is getting overstretched, and video media are following suit, expanding their physical and social boundaries. The early twenty-first century is witnessing a historical shift, whereby the moving image and the city are informing and redefining each other (Tweedie and Braester 2010). The prominent wall screens in Beijing collapse the distinction between environment and its representation, further complicating the material and symbolic spatial layering.

Qianmen: A Gate to the Future

The politics of emergence may be exemplified by the Qianmen Avenue reconstruction project, undertaken by Beijing's municipal government in 2007–2008 and later handed over to the private company SOHO China.[5] The avenue, leading to the Qianmen Gate and beyond it to Tiananmen Square, was remade into a pedestrian mall as part of Beijing's place-marketing strategy in preparation for the 2008 Olympic Games. Situated at the center of the capital, on the seam line between the touristy destinations around Tiananmen Square and the popular commercial venues of the old Outer City, the Qianmen construction project and the attendant imagery constituted a major spatial and visual intervention in the urban fabric. It is therefore a test case emblematic of the state of the image and the condition of visuality in Beijing under the aegis of the politics of emergence.

To be precise, it was not the Qianmen project that provided the rich symbolic imagery—the project was off-limits—but rather the walls that cordoned off the construction site. Since 2007, the area of about half a square kilometer had been enclosed by a high wall of metal plates, on which were displayed slogans and images, most prominently imaginary street-level and bird's-eye views of the district after the project's completion. (Although Qianmen Avenue was reopened in 2008, the project would take years to complete; in 2013, SOHO China announced that occupancy of the gentrified buildings reached 73 percent.[6]) The architectural simulations on the surrounding walls substituted, for the public eye, the material buildings erected behind the walls. The drawings ostensibly replicated what would be constructed, serving as temporal and spatial marks that addressed the visitor: You are standing at the right site but not at the right time. The image transports the viewer for a moment to the future. It corrects, so to speak, a temporal error, a miscalculation in time travel. The image is not just a wish fulfillment; it performs, rather, a necessary temporal readjustment. Insofar as the image anticipates its own destruction, when the wall is dismantled in the aftermath of the reconstruction, it is because of the assumption that the new buildings will align the viewing subject with the site's destiny.

A particular visual assemblage was presented on the eastern border of the reconstructed area. Qianmen East Road was demarcated by a decorative gray brick wall, which was in turn topped by a printed tarp featuring a panoramic view of Beijing in the early twentieth century (figure 1.1), a fifty-three-meter replica of Liu Hongkuan's twenty-one-meter scroll *The Palatial Capital* (*Tianqu danque*, completed in 2001). The scroll, which was also displayed at the Beijing Urban Planning Exhibition Hall, proceeds from the south along the Old City's central axis, passing by the Altar of Heaven, through Tiananmen, the Forbidden City, the Drum Tower, and out through the Andingmen Gate. The idealized image captures Beijing—then remade into a city of leisure (Dong 2003). Tinted with nostalgia for Beijing on the verge of modernization and for a vernacular architectural idiom, the scroll includes a depiction of the five-gate memorial arch (*pailou*) at Qianmen and the tramway line that went through it, long gone and now restored in the gentrified Qianmen Avenue. In yet another variation of replicating the image within through an image on the wall, the current restoration is fashioned as a historical return. The new Beijing is achieved by reattaching significance to vestiges and bringing them to a renewed flourish.

The Qianmen project seems, on its face, to encourage historical consciousness through engaged comparison between past and present. Yet, rather than permitting citizens to access the construction site, where they might actively

1.1. Workers mount a blown-up replica of Liu Hongkuan's scroll, *The Palatial Capital* (*Tianqu danque*, 2001, ink on paper) on top of the wall surrounding the Qianmen project. Still from *Qian Men Qian: A Disappearance Foretold* (2008; dir. Olivier Meys and Zhang Yaxuan).

engage in extricating the site's meanings, the Qianmen project instead presents the visitors with mediated views and preprocessed experiences. The new buildings at Qianmen lay claim to absolute transparency: They are ostensibly not vestiges of earlier architecture but rather fully and accurately restored replicas. Whereas for Certeau the palimpsest provides an opacity that "escap[es] the imaginary totalizations produced by the eye," the gentrified Qianmen is a ready-made product that can be taken in with a click of the camera before moving on without delay (de Certeau 1984: 93). Indeed, like the Tiananmen Gate just to its north, Qianmen Avenue has become a site for photo ops for tourists as a testimony to their visit to Beijing: in Joseph Brodsky's aphorism, "Kodak, ergo sum" (Boym 2001: 296). Such ritual photo taking imagines future viewing while asserting one's presence at the site (Braester 2010b). Qianmen invites and caters to the touristy gaze, a Disneyfied version of urban experience. The old Qianmen district is turned into a fetish, obviating the gap between past and present and legitimating the site's gentrification into a pedestrian shopping mall. The aporia of images, replicating the site on boards and screens on its perimeter, does not encourage historical distance but rather collapses space and time: The future is here and now.

Cinematic Panoramas and Apertures

The temporal overlaying at Qianmen was not only achieved through the installation of multiple visual references on the perimeter walls; it was also enhanced by the active participation of those who walked by the reconstruction site. Qianmen generated an experience based on photographic framing, cinematic panoramas, and interactive screens.[7] The Qianmen project exemplifies how the politics of emergence establishes a symbiosis with filmic presentation modes, in particular with twenty-first-century video and mobile screen culture.

The large panels surrounding the closed-off area, and especially the fifty-three-meter scroll depicting Republican-era Beijing, presume the gaze of an ambulatory visitor who takes in the substitute cityscape in motion. The scene unfolds in a quasi-linear fashion, leading the viewer through a paced, panoramic exposition. Liu Hongkuan's ink painting follows the structure of scrolls depicting processions approaching the Imperial Palace along Beijing's central axis, notably the seventeenth-century work *The Kangxi Emperor Returning to the Forbidden City from His Southern Expedition*.[8] Yet the contemporary painting lacks a political and narrative anchor of the kind that the imperial procession lends to the earlier scroll. The urban spectacle in Liu's scroll is more evenly distributed, dwelling on everyday life, akin to the famous *Along the River during the Qingming Festival* (*Qingming shanghe tu*, eleventh century). *The Palatial Capital* has indeed been called "an *Along the River during the Qingming Festival* of the twenty-first century." Furthermore, the intimacy of the original hand scroll—which was designed to be viewed in a collector's studio, where it would be carefully unrolled one section at a time—is lost in the blown-up version hung on top of the perimeter wall at Qianmen. It becomes a grand display, a large screen that has less to do with traditional ink painting and more in common with the silver screen. (Liu's scroll was later replaced by *Along the River during the Qingming Festival*, in what may be a reference to the latter's use in an enlarged, digitally animated version at the Shanghai Expo of 2010; scroll and screen literally became one.)

The piecemeal observation as the visitor to the site walks along the scroll turns the stationary panels, in effect, into moving images. The visitors are not simply confronted with the projected future. Instead, they are led to it in a process that creates wonder, expectation, and a sense of transcending one's physical presence. The images and screens unfurl and stretch. They cannot be viewed all at once, but in segments. They require regimented viewing— peeping in one section at a time, walking along them, or panning across them. The images of the future are not laid out statically in front of one another, to be seen from a single angle. Rather, the palimpsest is created and reaffirmed by the viewer's mobile participation.

The process of discovery, along a cinematic logic of images unfolding in time, was taken a step further at the passersby's initiative. A particular spectacle developed on the northern wall of the construction site, directly in front of the entrance to Qianmen Avenue. There, life-size realistic drawings on the surrounding temporary barrier, as well as a slideshow on a large LED screen above, provided the same views that would be available when the wall was removed (to be precise, the wall faced south, and the painted panels showed views facing north; figure 1.2). Yet what fascinated the passersby was not the visualizations as much as the ability to look in: By July 2008, one could see the almost-finished project through the wall. For a short period, the two forms of the city, material and virtual, coexisted. It was the ability to see one through the other that drew a steady stream of gawkers.

The opening looking into the construction area satisfied a voyeuristic desire. It was a rectangular hole, just above average eye level, about one foot high and three feet wide. Crowds gathered around this makeshift window, striving

1.2. Painted barrier at the Qianmen reconstruction site.
Photograph by the author.

to look into the almost-completed site and take photos. The opening must have been crafted by impatient citizens: The double canvas in front, printed with architectural drawings, was pulled down, and the metal sheet was folded down as well (figures 1.3, 1.4). The dozens of guards and policemen patrolling the area did not pay any attention. The impromptu action of peeping in seems to have coalesced with the authorities' interest in providing a preview of the emerging gentrified district.

By tearing out a rectangular window, roughly the same size as a high-definition TV, the visitors framed the future scenery as if it were a video screen. The gesture was replicated when the crowd proceeded to take photos through the window, aligning the opening with their mobile devices. Since

1.3 and 1.4. Onlookers peering through the painted barrier at the Qianmen reconstruction site. Photographs by the author.

many were too short to look directly through the window, they were instead viewing the avenue vicariously on camera screens, using them like periscopes. The citizens looked no farther than the cameras, which overlapped with the wall. The camera screens function less as windows than as screen walls that replicated the view beyond. The overlaying of the present and future Qianmen was achieved through multiple screenlike apertures, rendering the urban palimpsest coterminous with the proliferation of screens.

The image making at Qianmen exemplifies also a growing reliance on the mobility of screens. The digital skin around buildings no longer has to coincide with the fixed architectural structure. It can also be generated by screens placed around it at the spur of the moment. With the ubiquity of mobile screens, in the form of digital cameras and smartphones, the wall screen can be outsourced to the consumers and replaced by mobile devices. The symbiosis between architecture and digital imaging allows for the construction of ghost temporalities, fake façades, and spurious visual archives.

Some viewers had with them regular digital cameras, suggesting that they were either tourists or locals arriving at the site with a premeditated intent to take photos. Others were using smartphones, which at the time were still a novelty (the iPhone was released in June 2007, and phones running Windows Mobile had become popular just before that). These viewers, very likely passersby who joined the spectacle with the nonchalance of youngsters whipping out their mobile devices, illustrate a new manner of consuming the urban palimpsest. Attracted by a casual glance, these amateur photographers produce, store, and disseminate images with facility through mobile networks. Insofar as the politics of emergence is founded on enunciating a temporal disjunction—"you are standing at the right site but not at the right time"—the networked smartphone signals a new form of spatial and temporal overlaying. The practically infinite capacity to store, retrieve, and share digital images and videos in real time results in the phone holder's ability to say "I was here," about any time and at any time. The palimpsest can be invoked and modified with the single push of a button.

The visual duplication of the site, this time instigated by citizens using their personal screens, might seem to establish a different power structure from that established by the large billboards erected by the municipal government. Yet using this new manner of producing and consuming the urban palimpsest amounts to the privatization of the spectacle in the age of interactive communication. As mobile screens are taken along and used for photographing various sites, they create aggregate cityscapes. Helen Grace (2007) draws a sharp

1.5. LED screen on the perimeter wall around the Qianmen reconstruction site. Photograph by the author.

distinction between imposing urban architecture and user-generated content such as cellphone photos. For Grace, these photos are "an archive of the present." They, and the commonplace structures they document, are exponents of "horizontal monumentality" (468). Like a monument, they generate a ritual of commemoration, yet they do so without resorting to the overwhelming verticality of high rises. I submit that, departing from Grace's view, even such archives of the present can pose an impediment to historical thinking. Insofar as the photoscapes are monuments, they tend to be monuments of the present alone. Like the billboards and large-screen video displays, the mobile screens facilitate a radical presentism in the guise of paying tribute to the history of sites.

Even though then-nascent networked screen technologies were not employed in any obvious manner at Qianmen at the time observed, their influence on temporal perception was felt in a number of significant ways. Screens were integrated into the cityscape—not only the jumbotron at Tiananmen Square just to the Qianmen's north, but also the aforementioned screen above the perimeter wall. The slideshow transitions, including pixelated dissolves and wheel wipes, had the look of PowerPoint presentations (figure 1.5). The

TRACES OF THE FUTURE 31

overlaid views of the city employed the visual vocabulary of computerized data, of the same kind that visitors would be using on their personal computers. (I was also present when the computer running the slideshow crashed; after the blue screen of death, the screen showed the reboot code running, stressing the familiarity of home media.) Moreover, the LED display and the canvas panels alluded to a digitized, interactive environment. The drawings, some of which were produced with architectural simulation software and bore resemblance to images used in real estate sales promotion, showed the reconstructed site populated with many pedestrians, uncannily suspended in motion. The panels aspired to the aesthetics of animated simulation and augmented reality, envisioning the emerging material constructions as if they were a video game and rendering the urban palimpsest reliant on virtual technologies.

Hypervisual Space, or, Taking Metaphors Literally

The urban palimpsest instates an anticipatory visual regime, idealizing the city in the form of its potential future. The resulting politics of emergence helps explain the proliferation of some images in China's cities and the absence of others. I have noted elsewhere Beijing's and Shanghai's recent inundations with drawings, maquettes, and videos of utopian and dystopic architecture (Braester 2013). These and other Chinese cities are defined by an abundance of architectural representations—made, displayed, and replicated at a variety of sites and in many types of media. This hypervisuality of space betrays the developers' anxiety about the efficacy of images: Perhaps throwing more and more of these images at citizens could induce the desired change in the prevailing spatial perception. The hegemonic discourse has been abetted by a visual campaign of shock-and-awe.

At the same time, the public images convey an interdiction. The high walls at Qianmen blocked outside viewers from witnessing the contentious demolition-and-relocation. The gentrification of Qianmen, like many other demolition-and-relocation projects around Beijing, was not welcomed by all the area's residents. Many felt that they were evacuated forcefully, destroying their community and livelihood, and given a pittance for compensation. To go beyond the images on the surrounding walls one would have to trespass into the off-limits construction site. Police stood on guard to check any person walking into the area. Imagining anything in excess of the future that the images readily suggest would be a transgression according to the municipal authorities. Making other, alternative images would deliver a statement against the politics of emergence. In the case of the Qianmen gentrification project, it took a dedicated effort by independent filmmakers to go behind the wall and bring images to light from where a ruthless uprooting of the near-indigent population was under way. The documentaries *Meishi Street* (dir. Ou Ning, 2006) and *Qian Men Qian: A Disappearance Foretold* (2008; dir. Olivier Meys and Zhang Yaxuan) describe the demolition-and-relocation project from the viewpoint of residents, who speak up and expose the injustice done to them. The latter film shows a behind-the-scenes portrait of the Qianmen façades. Tan Tan's video artwork *Qianmen Hotel No. 1* (2008), which traces the spaces of a now-evicted guesthouse, provides a mirror-image view from the bordering perimeter. Tan Tan looks out of the reconstruction site, from a building designated for demolition. The video returns the gaze peeping in and shows the passage of time, the accumulation of dust, and the restless ghosts of the past. All three works avail themselves of the trope of excavation but keep their lenses trained on the present. The spatial transformation is visualized in real time. Digital video—the guerrilla weapon of choice against the extravagant, expensive, and high-tech media deployed by developers—points critically at the radical temporal compression and the creation of a uniform, globalized space.

The few counterexamples notwithstanding, the visual strategies at Qianmen demonstrate that new media mostly support the logic of the palimpsest. The degenerate utopia of the gentrified pedestrian mall is complemented by an unsustainable techno-utopia. More specifically, the billboards' allusion to augmented reality, and their appeal to visitors' mobile sensorium, regulate the viewers' experience, defying any unconditionally salutary approach to new media (as advanced for example by Lev Manovich [2001]). The superimposition of the human eye, the camera viewfinder, the digital camera screen, and the wall have produced a complicity of imaging practices with spatial and temporal manipulation. The multiplicity of apparatuses has framed the citizen within a Chinese box of replicated images.

The Qianmen project demonstrates how the superimposition of media avoids prioritizing one image over the other, resulting in a hall of mirrors that defers any critical view. Contemporary screen practices tend to support the absence of representational hierarchy. Greg Siegel (2002) has observed the function of large-scale video displays in sports stadiums. He notes how such screens, instead of contrasting the digital image and the event, operate within a continuum that includes the profilmic and the mediated. Viewers shuttle back and forth between the unmediated event and the onscreen image without privileging either. Likewise, the billboards, peeping holes, and mobile screens at Qianmen present a layered spectacle—yet one that uniformly supports the hegemonic view. The anticipatory visual rhetoric of the politics of emergence relies on the hypervisual multiplication of images and on ignoring how its various components might be at odds with one another.

The aporia of images at sites such as Qianmen might be identified too readily as a free-for-all simulacrum. The trope of the palimpsest proposes temporal immediacy and spatial transparency, yet the anxiety underlying hypervisuality manifests itself not in abstraction but rather in taking images literally. The palimpsest avows that the utopian promise of architecture is eminently visualizable. A recent billboard seen on Chinese streets has made it into the news. It advertises plastic surgery by pledging "a nose as pretty as the Eiffel Tower" or "a pretty nose resembling the Eiffel Tower" (*Aifeier meibi*).⁹ Ignoring the warning in Jia Zhangke's *The World* (2004), where a Chinese replica of the same tower is featured to warn that architectural façades conceal a structure of economic inequality, the rhetoric of the palimpsest and the politics of emergence conflate architecture and image, to the point where all screens melt into air.

Notes

1 In Jewish rabbinic literature, redemption is signaled by God's entry into a heavenly temple in Celestial Jerusalem, which must be preceded by the reconstruction of Earthly Jerusalem. The idea of Jerusalem as a doubled site is owed to the Babylonian belief in parallel universes: A heavenly replica exists of the entire earthly universe (Tsafrir, Safrai, and Stern 1999: 48).
2 Edward Soja (2000: 337–339) squarely criticizes Boyer for an inflexible, dystopic vision.
3 For mentions of the palimpsest, see, e.g., HUTOPOLIS (www.hutopolis.info/) and WAI Architecture Think Tank (http://waiarchitecture.blogspot.com/2012/03/what-about-ideal-cities-and-counter.html).
4 Paul Virilio, "Architecture in the Age of Its Virtual Disappearance" (interview, 1993). Quoted in Friedberg (2004: 183).
5 See China SOHO's site for Qianmen's development at www.qianmensoho.com/en.
6 See "SOHO China Announces 2013 Annual Results," SOHO China, March 4, 2014, http://www.sohochina.com/en/news-media/43387.

7 For a discussion of panoramic views as predecessors to the logic of contemporary mobile screens, see Verhoeff (2012).
8 On premodern scrolls of Beijing, and the Kangxi scroll in particular, see Wu Hung (2005: 167–171).
9 Joshua Melvin, "'Eiffel Tower' Nose Surgery Booms in China," www.thelocal.fr/20140130/eiffel tower-nose-surgery-booms-in-china.

2

THE CHINESE ECO-CITY AND SUBURBANIZATION PLANNING

Case Studies of Tongzhou, Lingang, and Dujiangyan

Robin Visser

If we plan it, they will come.
—Yang Xifeng, vice director, Lingang New City Administrative Council (2006)

When you visit Thames Town of Songjiang New City ... it feels like a ghost city.
—Wang Lan, Ratoola Kundu, and Xiangming Chen, "Building What and Whom?"

In 1970—in his first sustained critique of urbanism, *The Urban Revolution*—Henri Lefebvre denounced urbanism as the vehicle for a form of rationality in which "space, deceptively neutral and apolitical, constitutes an object" (2003: 164). While the production of space is not new in itself, he argued, the production of social space on a global scale results in the "urban illusion," replacing the "state illusion" (and before that, the "philosophical illusion") as the latest totalizing ideology. This neoliberal ideology arose because "capitalism appears to be out of steam. It found new inspiration in the conquest of space—in trivial terms, real estate speculation, capital projects (inside and outside the city), the buying and selling of space. And it did so on a worldwide scale" (2003: 155).

In this chapter I examine Lefebvre's notion of the urban illusion by examining three case studies of eco-city development in China. Lefebvre's critique of urbanism rejected not only the role and strategy of the state in spatial politics, but also the leftist idea that the "Chinese way" offered an alternative to global urbanism. In his words: "In China today, like the Soviet Union yesterday, cities continue to grow along with the economy, and possibly, the increase in speed. As they do elsewhere.... The urban revolution is a planetary phenomenon.... Moreover, if the 'global city' is of interest to the theoreticians of the 'Chinese way,' the eventual 'suburbanization' of a large portion of the world is of no less interest to urban strategy" (2003: 113). Lefebvre refers to the "Chinese way" as "suburbanization" due to Maoist ideals aimed at leveling class distinctions between urban and rural. These ideals were only partially enacted under Mao, first in the building of socialist workers' villages and decentralized "scattered collectives" in urban fringe areas (1952–1959), and then in mandated relocations between rural and urban locales during the Cultural Revolution (1966–1976).

From a twenty-first-century vantage point, Lefebvre appears prescient in anticipating how "suburbanization" and an "increase in speed" would prefigure global urban trends. In fact, although technically still referred to as an "urban revolution,"[1] most urban growth today takes the form of suburban (sometimes also referred to as peripheral, informal, or peri-urban) development.[2] The Chinese term for "suburb," *jiaoqu*, merely denotes a locale near the city rather than a distinctive settlement, yet such locales are undergoing dramatic spatial restructuring via "new city" planning. In this chapter, I define suburban development as the transformation of suburban land into more intensive industrial, commercial, and housing developments, regardless of where the impetus for such changes comes from. The forces driving suburban development may be centrifugal in nature (originating from the central city) or centripetal (coming from more distant places owing to the attractiveness of the central city and its suburban areas) (Zhou and Ma 2000: 206). By the mid-2000s, strategies of (sub)urbanization planning in China as drivers of economic growth had largely usurped practices of urban planning based on physical forms (Kang 2006). The Ministry of Construction changed its name to the Ministry of Housing and Urban-Rural Development, with planning governed by the 2007 Urban-Rural Planning Law (replacing the 1989 Urban Planning Law). Under the banner of its 2007 central policy of regional "urban-rural integration" (*chengxiang tongchou*), China now leads what might be called the worldwide "suburbanization revolution."

Where Lefebvre's analysis of future urban trends fell short was in his assumption that the state was no longer a central player in the "urban illusion." As Ananya Roy points out, due to Lefebvre's legacy, "there is a tendency to imagine the 'informal' as a sphere of unregulated, even illegal, activity, outside the scope of the state, a domain of survival by the poor and marginalized, often wiped out by gentrification and redevelopment" (2009: 825–826). Yet, as she argues elsewhere (Roy and AlSayyad 2003), this discourse masks the fact that informality operates within the purview of the state and that the state itself operates informally to flexibly access territory, while bypassing formal mechanisms of accumulation and legitimation. Nick Smith (2014) draws similar conclusions in his recent case study of Chongqing peri-urban development. His findings can likely be generalized given that an estimated 60 percent of urban development in China is "informal" and based on "illegal" land use (Li 2012). Shen and Wu (2013) explain how rural-to-urban land transfers in suburbs are facilitated by government actors:

> The process supporting both industrial and housing development on the urban edge is rapid urbanization facilitated by strong state intervention. Since a massive amount of land is needed for rapid urbanization, local entrepreneurial governments often set up a series of institutions, such as land circulation, housing plot exchange policy, or land for social insurance policy. In most cases the district government is able to acquire collectively owned land by offering compensation packages to the collective authorities and farmers. (1827)

By presenting three case studies that trace the past decade of shifting policies promoting new town development aimed at integrating vast metropolitan regions, I identify mechanisms of rural land conversion in China's ever-evolving suburbanization strategies. The People's Republic of China currently has a two-tiered land system based on China's Land Administration Law and Urban Real Estate Development and Management Law: Urban land is owned by the state and rural land is mostly owned by village collectives. In my analysis I attend to how the rhetoric of eco-city development, as just one of several new town branding trends over the past decade, rationalizes rural land transfers. Although China has promoted sustainable development since the mid-2000s, particularly in the form of new urbanism ideals in transit-oriented development (TOD) satellite towns (Duan and Yin 2011; Tang 2008), most experts indicate these satellite towns "ultimately fail to meet their [sustainability] goals" (Tan 2010: 47). Instead, entrepreneurial local governments and institutions skillfully utilize their power to convert farmlands to constructed lands for various kinds of industrial and commercial development. The primary value of most of these projects lies in their position within a speculative economy rather than their direct contribution to the social or ecological good. Progressive urbanism, as such, is an illusion.

The first case analyzes the rhetoric, plans, and development of Lingang ("Marine Eco-City") in Shanghai Municipality, which I visited with urban planning academics in December 2006. The second analyzes the transformation of Tongzhou ("Low-Carbon Metropolis"), one of Beijing's eleven satellite towns, which I first visited in May 2007 and revisited in November 2015. The third examines the rationale for the postearthquake reconstruction plans for Dujiangyan ("Ecological Garden City") in Sichuan Province, presented in July 2008 to an international delegation of urban planning and disaster reconstruction academics. Despite regional differences, what is fascinating is how the rhetoric of TOD morphed into "eco-city" planning and, most recently, to "integrated regional development," according to shifting state priorities and policy incentives.[3]

Each case challenges the prevailing hypothesis that urban developmental policies are primarily driven by what the Chinese state avows are necessary measures to resolve economic and planning exigencies resulting from the two-tiered land system. Instead, I consider the thesis that the urban-rural divide is an institutional regime that, as Nick Smith puts it, is "both resilient and socially efficacious, but is nonetheless artificial and thus prone to re-appropriation" (2014: 373). The artificiality of the constructs of "urban" and "rural" can, I surmise, facilitate the state in sustaining economic growth by incentivizing entrepreneurial local governments to adopt flexible land development strategies (Chien 2013; Kang 2006). Second, I examine the hypothesis that new city plans manifest as isomorphic forms independent of heritage, scale, or local conditions (Gaubatz 2008). A related question considers whether postspatial strategies adopted by developers have, in fact, preempted practices of urban and economic planning informed by physical or temporal data. It queries whether the avowed social and ecological aims of central government and municipal planners will be achieved, given the virtual catalysts of urban entrepreneurialism. Some scholars predict that ecologically sustainable techno-utopias, such as Lingang, foretell future global urban development in which the demise of street life is inevitable (Kim 2014). If so, are disembodied ghost cities sustainable forms of human technology over the long term?

Eco-Cities and Suburbanization Planning in Twenty-First-Century China

The prevalent form of rural land conversion in China since 1980 is the *xincheng*, variously translated as the "new city," "new town," or "satellite town." For the purposes of this chapter, this term is understood in its broadest sense, and it incorporates urban planning policies that have evolved through three distinct development phases and strategies, culminating with the contemporary new city development strategy of the eco-city. According to Duan Jin and Yin Ming, the first phase, which began in the 1980s, was the industrialized development zone, which was generally located within new districts of existing cities such as Dalian. This phase also included economic and technological development zones, science parks, and bond areas. As early as 1993, more than six thousand development zones occupied fifteen thousand square kilometers of rural land, exceeding the built-up area of all the cities in China (Hsing 2010: 105). The second phase, governed by an "urbanization development strategy," aimed at transforming an urban core into polynucleated forms, such as Guangzhou New Town. The third phase ushered in suburbanization planning strategies of regional "urban-rural integration," with an emphasis on low-carbon or eco-city development, such as Tianjin New Eco-City (Duan and Yin 2011: 182). As is clear from table 2.1, eco-city development in China is but the latest in several phases of new city development.

As governance became increasingly decentralized in the 1990s, economic planning (*jihua*) and urban planning (*guihua*) gave way to what, by the 2000s, was termed strategic planning (*cehua*), a term borrowed from business management and marketing. Real estate developers and marketing gurus advised mayors on directions of urban operation and management. Developers began calling themselves "businessmen of urban management," their aims clearly at odds with the ostensible functions of the twentieth-century discipline and profession of urban planning. Rather than simply zoning existing parcels of land, planning was increasingly approached more proactively by strategizing new *xincheng* projects according to principles of place production, place consumption, and place marketing. You-tien Hsing characterizes the new approach to planning as a type of urban entrepreneurialism with a clear purpose to "create [property] value in the city" (2010: 104). Such entrepreneurialism by local municipalities is facilitated by central government initiatives, yet new city planning is often implemented in ways that run counter to central policy aims. Socialist planning legacies, coupled with current economic and political exigencies, result in the rapid building of new cities radically dissociated from their regional ecology and local heritage through redeveloping and expanding old cities or occupying major portions of land in the remote countryside.

After Minister of Culture Doje Cering announced in 2000 that four hundred new cities would be built by 2020, the construction of new cities began in earnest across China's vast territory. The "Go West" campaign fostered the

Table 2.1 Three Stages of New City Development in Post-Mao China

	First-Generation New Cities	Second-Generation New Cities	Third-Generation New Cities: Prospects
Regional Development Strategy	Industrialization drives urbanization	Urbanization development strategy	Urbanization development strategy
Urban Development Policy	Economic and technological development zone	Expand cities and organically evacuate old towns	Urban-rural integration, integrated development of district cities
Urban Spatial Structure	Mononucleated	Transition from mononucleated to polynucleated	Polynucleated within cities, networked between cities
Construction Aim	Economic growth	Comprehensive metropolis	Fully adopt eco-technology; construct sustainably developed model district cities
Siting Principles	Cost priorities	Location priorities	Areas that cannot be utilized under traditional conditions, such as salt flats, deserts, etc.
Initiation Mode	Industrial initiatives	Catalyst-driven	Sustainable geographical priorities
Representative New Cities	Dalian Economic Technology Development Zone	Guangzhou New Town	Tianjin New Eco-City

Source: Adapted from chart 6.2 in Duan and Yin 2011: 182.

development of new cities in the interior provinces, while the strategy of land-sea coordinated development is a central component of the Twelfth Five-Year Plan (2011–2015). Disaster recovery, such as from the 2008 Wenchuan earthquake, provided more opportunities for new city development, while large municipalities expanded major portions of their territory. These new cities are dubbed "university cities," "auto cities," "garden cities," "tourist cities," "corporate parks," "CBDs," "eco-cities," and so forth. Ranging in size from 50 to 350 square kilometers, they include the new seaside Tianjin district, Shenbei in Shenyang, the Olympic new city in Nanjing, Lingang New City in Shanghai Municipality, the new waterfront in Hangzhou, and Ordos in Inner Mongolia.

Such cities are now branded by states as commodities in a global supply chain. For example, when Sweden marketed the "Sustainable City" at the 2002 World Summit of Sustainable Development in Johannesburg, it attracted a great deal of attention from the Chinese government. By 2008 at the World Urban Forum in Nanjing, the two nations had formed a Sino-Swedish Environmental Technology Cooperation that demonstrated a Sino-Swedish Low-Carbon Eco-City at the 2010 World Expo in Shanghai (Hult 2013). Chinese central government initiatives were, in turn, quickly enacted locally. Catalyzed by policies promoting "ecological civilization," local governments started planning to construct eco-cities, low-carbon cities, or sustainable city projects. A study done by the Chinese City Science Association in 2011 reported that more than 80 percent of prefecture-level cities had at least one eco-city project, and as of 2012 around two hundred eco-city projects were proposed, under construction, or implemented (Chien 2013), and twenty-one marine eco-cities were planned on reclaimed land on China's eastern seaboard (Lin et al. 2011). The *China Real Estate Post* (April 2012) makes it clear that eco-cities are merely speculative opportunities: "the 'low carbon eco-city' strategy has become the third wave of local development fever, following the first wave of development zone fever and the second wave of college town fever over the past three decades, becoming a dream of almost every prefecture-level city."

Scholars such as Chien Shiuh-Shen contend that this third phase of new city development has very little to do with ecological and social sustainability, but is the latest strategy for the neoliberal production of space that has dominated local entrepreneurialism throughout the post-Mao era. He argues:

> This assessment shows that the core characteristic of land conversion has remained unchanged. Local authorities skillfully apply central land regulations in order to secure additional local land finance. Adopting the new

discourses behind development zones in the 1980s, college towns in the late 1990s, and eco-cities in the mid-2000s, it is true that within the new "wineskin" (that is, a new development concept) is still the same old wine (entrepreneurial land development projects). Beneath the image makeover and city marketing lie corporate investment, duplicate constructions, industrial isomorphism, illegal land appropriation, and real estate development—fundamental challenges confronting Chinese local entrepreneurialism over the past three decades. (2013: 186)

In other words, rather than actually developing a knowledge economy or green economy, such land-related projects are strategies used by Chinese subnational authorities, in collusion with transnational corporations, to generate local finance and real estate development. Similarly, Xuewen Tan argues that new town development is merely a speculative urbanization strategy rather than a public good in itself. He writes:

New towns or satellite towns, strictly defined, do not exist in China, nor do new town policies, although satellite towns and new towns can be seen everywhere on China's city planning and construction drawing boards. The so-called satellite towns or new towns are just labels on one tier within a city system. Since there were no genuine new-town policies, the benefit to those who migrated from other areas due to satellite town construction cannot be seen as "externalities" of regional public good. It is mainly the result of urbanization. (2010: 58)

Both critics insinuate that post-Mao urban development falls in the category of Lefebvre's spectral urbanism—that is, the virtual function of urbanism as a totalizing ideology is to facilitate the circulation of capital rather than to create a genuine public good.

Lingang New City and Shanghai "1966"

Urban planning in modern China has emphasized suburbanization, and Shanghai has been no exception. The Great Shanghai Plan in 1946 followed the idea of "organic decentralization" and proposed a series of townships, each of which would aim to have 160,000 to 180,000 people in the suburban area of Shanghai, with the new towns to be planned under the guidance of the Garden City model developed by Ebenezer Howard. Twelve satellite towns were identified in the 1959 Master Plan to accelerate industrialization and accommodate workers in the factories. In the twenty-first century, however, Shanghai's new stage of new town suburbanization has focused on real estate development rather than industrialization. In 2006 Shanghai unveiled a regional urban system plan called "1966," which stands for one central city, nine new towns, sixty new townships, and six hundred central villages spread throughout the entire municipal region (L. Wang et al. 2010).

Among the nine new towns of Shanghai, Lingang New City has an exceptionally utopian layout, with water playing a pivotal role in the overall urban scheme. Lingang New City is built on reclaimed land sixty kilometers south of Shanghai for an estimated population of eight hundred thousand. Located at the southeast tip of Shanghai Municipality in Nanhui district, it aspires to be a worthy complement to its neighboring mega projects, the Yangshan Deepwater Port and the East China Sea Bridge, the latter of which a press release claims is "believed to be one of the two largest man-made constructions in the world today, the other being the Three Gorges Dam."[4] Built on a landfill of seventy-four square kilometers, Lingang is situated at a strategic geographical point, an outward-focused "space of flow" that connects nautical, airborne, and terrestrial passages (the A30 highway is an hour's drive to Pudong International Airport, and the planned extension of the Number 11 subway line will take commuters back and forth to metropolitan Shanghai).

In 2002 the Pudong New District of Shanghai Municipality, which initially oversaw the Lingang New City development, conducted an international competition and selected the German design firm Von Gerkan Marg und Partner (GMP), well known for its architectural modernism. Lingang New City is inspired by the image of a drop of water hitting the center of a huge manmade lake, with the resulting ripples fanning out into the built environment of the surrounding city, creating the paradoxical image of an "empty" city center. The city radiates outward from the lake in three bands of functional zones separated by ring roads. The innermost band is the commercial district; the middle band is the green belt, planned for recreational and civic functions; and the outermost band contains twenty-four residential "island" clusters. Boulevards extend from the lake in the eight cardinal directions, intersecting the ring roads and fanning the circular theme outward (figure 2.1). Each residential cluster is separated by villas and greens, with canals extending from Dishui Lake to form a blue network running through each residential island.

As I mention in my book *Cities Surround the Countryside* (2010), I first visited Lingang New City in 2006 while the reclaimed land was still marshy. Mr. Yang Xifeng, the vice director of Lingang New City Administrative Council,

2.1. Concept plan for Lingang New City on display.
Photo courtesy of Jungin Kim.

gave us a tour of the site and showed us conceptual sketches and planning layouts of the city. The administrators took evident pride in the city's utopian layout, and Dishui Lake was said to cover the exact area of the famed West Lake in Hangzhou. Yet despite the fact that the city was being promoted as a center of education, tourism, and culture, visiting academics conveyed skepticism about these functions. When a Fudan University doctoral candidate in urban planning exclaimed that educated people would refuse to live in such a remote cultural backwater, Mr. Yang conceded that most residents are likely to be former agricultural workers reemployed by industries supporting the Yangshan Deepwater Port.

The question of who will inhabit new cities such as Lingang is pertinent. In Jung In Kim's account of Lingang after a visit in 2012, he says,

> It projects deadly silence in space . . . the feeling of estrangement by visitors is amplified when they see abandoned constructions and meager portions of the New Bund completed. Most of the lots around the lake remain empty, creating a barren landscape and revealing the city's financial troubles. The

2.2. Uninhabited residential clusters in Lingang in 2012.
Source: Kim (2014: 346).

> city thus took a virtual turn where it seems to be a bizarre stage without its citizens, only occupied by a small group of sojourners and groups of tourists near the lake. (2014: 345–346)

A subheading in a book chapter on Lingang is blunt: "Key Challenge: A City with Physical Development but No People" (Sha et al. 2014: 136) (figure 2.2). That the Lingang New City project is problematic is clear from the fact that since its inception in 2002, its administrative management structure has changed four times (Li Zhen 2014). From 2005 to 2009, Lingang New District also incorporated Fengxian New District into its regional planning; as a result, many funds aimed at the Lingang New City development were diverted into other projects in the two districts. In 2010–2013 another administrative phase occurred where the city's name changed from Lingang Marine Eco-City to Nanhui New City. The March 1, 2013, edition of the *Shanghai Financial News* reported Shanghai Municipality's injection of an additional one billion yuan into Lingang New City after yet another administrative restructuring promoting regional integration, and, to attract potential residents or investors, a slashing of going real estate prices by half to 6,000 yuan per square meter.

To date, the land reclamation has been completed along with the completion of Dishui Lake. Much of the road infrastructure has already been built, in-

2.3. The Maritime Museum was one of the first buildings to be realized in Lingang New City, a town built on water. Photo: Hans-Georg Esch; source: "Maritime Museum, Lingang New City," *Steel Construction* 3, no. 2 (2010).

cluding the ring and radiating roads and the major thoroughfares to Shanghai. Likewise, most of the administrative and landmark attraction buildings, such as the Maritime Museum (figure 2.3), are occupied. Yet the two planned metro stations are yet to be completed, so access to the city is difficult for lower-income visitors or potential residents. In fact, the buildings and infrastructure that have been realized primarily showcase accomplishments to outside visitors rather than facilitate residential functionality. Notably, this phenomenon is not unique to China. As Jung In Kim argues in his analysis of three eco-cities (Lingang in Shanghai, Yijiapu in Tianjin, and Songdo in Seoul),

> Functional disparity and organizational rationality are frequently framed by a more dogmatic and political agenda of national competitiveness. In all three cities, spatial arrangement and positioning are a priori determinants, not because they benefit the everyday welfare of the citizens, but because they are expected to boost the national economy from a macro-economic perspective. The privileging of the nation to determine the strategic organization of

space and the positioning of high-skill industries and transportation hubs around the new city expound the perception that state-of-the-art territorial strategy means equating city building with the visible manifestation of national development expressed in the languages of "the first," "the largest" and "the most impressive." (2014: 348)

Many of Shanghai's other nine "new towns" also remain "ghost towns" to this day (Jin 2014). A 2010 description of Songjiang New City still resonates:

> When you visit Thames Town of Songjiang New City, you are likely to see few people on the streets. The first thing you will notice upon entering are the lovely guards in redcoats. It feels like a ghost city, but functions as a weekend resort or a popular scene for wedding photos, especially in front of the town's central church towering over the Tudor-style single-family houses and the British-style pubs nearby. The vacancy in Thames Town, however, does not mean a low purchase rate of its expensive houses, which have been largely sold out. The scarcity of low-density single-family houses with plenty of green space brings value and investment potential to Thames Town. (L. Wang et al. 2010: 337)

Because there are limited opportunities for investment in China, "hot money" chases high-profit housing in and around Shanghai and other large Chinese cities. Yet because there is no further investment in new town infrastructure, luxury residential environments function more as virtual realities than as actual inhabited communities. In this sense, twenty-first-century urbanism may be an unprecedented phenomenon. Fulong Wu, a Chinese urbanism expert, has argued that the sole reason rural lands are converted for urban development is because "the city is used as a fix to absorb capital" (2007: 9). As a result, although semipublic developers provide start-up funds for the initial transportation infrastructure, the spin-off projects created by entrepreneurial local officials generate significant personal profits but few public goods such as health care, employment, sustainability, or local infrastructure for inhabitants of new towns.

Tongzhou New City and Beijing 2020

In the 2004 Master Plan of Beijing (2004–2020), Tongzhou New Town is the largest (forty-two square kilometers) of the eleven key towns around Beijing. Emblematic of suburbanization planning strategies, it has recently been promoted to being an "International New City" from its previous status as a "Key New Town." Unlike Lingang, Tongzhou is not, in fact, a "new" city but rather is an ancient settlement established in the Western Han dynasty (206 BCE–24 AD). In the 1950s, Tongzhou was designated an industrial "satellite town" administered by Beijing, when dozens of state-owned factories were relocated there. With land market reforms in 1992, its real estate market slowly activated and the Beijing 2020 Plan identified Tongzhou as one of three key growth areas after the 2008 Beijing Olympics. Unlike Lingang, the municipal transportation infrastructure facilitates easy access to the city center. Located twelve miles east of Beijing's Central Business District (hence dubbed the "back garden of the CBD") (figure 2.4), one can easily access Beijing via Metro Line 8 and the Beijing-Tongzhou Highway. In 2011 Luo Tingting and Gao Xiaowei observed, "this leads to a problem in that there are houses but no businesses in Tongzhou" (2011: 503). Yet unlike Lingang New City, which still lacks urban vibrancy in late 2015, Tongzhou New City appears poised to explode. In June 2015 the Beijing municipal government made public its plans to move from central Beijing to Tongzhou New City. Many public hospitals, universities (including a new campus for Renmin University), and large state-owned enterprises will move as well, amounting to what the *China Daily* describes as "a shift of people equivalent to 15 percent of the city's population of over 21 million" (Zheng and Yang 2015).

When I visited Tongzhou in 2007 to interview the performance artist Zhu Ming, who had moved to Beijing from rural Anhui in 1992, he was paying RMB 800 as a monthly rent for a spacious apartment in a typical six-story Soviet-style apartment complex (figure 2.5). The dusty city had not yet become the latest development hot spot, and Zhu Ming could easily walk to the open-air market to buy fresh plants, vegetables, and live fish for lunch. A nearby medical clinic, school, and local bus system provided adequate infrastructure, while the subway into Beijing allowed him relatively easy access to the city when he needed it. Zhu Ming's lifestyle matched the claims for Tongzhou New Town in the Beijing 2020 Master Plan, boldly promoted in display rooms devoted to each of Beijing's eleven new towns during the opening of the Beijing Planning Exhibition Hall in 2005. Each new town was branded to attract residents and investments based on particular functions and tastes: "Shunyi is designed primarily for industrial workers," "Yizhuang will accommodate white-collar employees," and "Tongzhou welcomes migrant artists."

One of the reasons why the Tongzhou township government was eager to participate in such branding was that the township could profit from illegal

2.4. Advertisement for low-density housing in Tongzhou, "back garden of the CBD." Photo by author, 2007.

2.5. Zhu Ming (a migrant artist from Anhui) in a spacious, affordable Tongzhou apartment. Photo by author, 2007.

kickbacks from construction and informal housing allocation privileges. Paik and Lee define "quasi-owned housing" (*xiao chanquan fang*) in the suburban land development market "as an illegal residential building, constructed on rural collective land in suburban areas by the joint efforts of low-level local governments, entrepreneurs, and peasants for selling and renting to non-local urbanites who seek cheaper housing outside the city" (2012: 262–263). Tongzhou encouraged artist urbanites to illegally buy and rent "quasi-owned housing" from local peasants by officially designating it as a "special art district" as sanctioned by the Beijing 2020 Master Plan. In turn, "this massive revenue helps revenue-starved local governments, enriches rent-seeking officials, and simultaneously creates a large boost to the local economy" (Paik and Lee 2012: 278).

Nonetheless, as real estate speculation began to gentrify Tongzhou, such strategies became ineffective. By 2008 a housing bubble in Beijing had made Tongzhou attractive to far more urban investors than starving artists (Li and Chand 2013), so Beijing Municipality sought more control over its development. The 2007 Urban-Rural Integration policy aimed to stem devolution of power to localities by returning regional development governance to provincial capitals such as Beijing (for Tongzhou), Chengdu (for Dujiangyan), and Shanghai (for Lingang). As land prices began to soar, industries were required to relocate. By 2008 over 86 percent of manufacturing plants were located outside of Beijing's Fourth Ring Road. This was ostensibly due to environmental concerns but was primarily due to impacts on Beijing Municipality revenue (Gao et al. 2014: 8).

In the "Beijing 2010 State of the Environment Report," Tongzhou was singled out as needing immediate remediation due to severe environmental degradation. The Beijing Municipality subsequently announced it would intensify Tongzhou development, vowing to "consolidate its strength and focus on Tongzhou" (Peng 2014: 168). Tongzhou was soon promoted as a "low-carbon international city," with its 2012 Master Plan touting it as using "new city planning methods focused around creating environments that are sustainable yet provide a multi-functional new city complete with cultural facilities and a strong industrial base."[5] (See figure 2.6.) The developers claim that Tongzhou "is expected to achieve a 20 percent carbon footprint reduction from current conditions, resulting in a low carbon metropolis which will set new standards for development worldwide."[6] Tongzhou's transformation from a new town to a low-carbon international city is aimed at financing its development as the eastern hub of the "Eastern Development Corridor," a vast regional transportation network specified in the Beijing 2020 plan. On June 29, 2013, the eleventh Beijing Municipal Party Congress announced the development of Tongzhou New City as "the sub-center of Beijing" and less than thirty months later, the former residents of the Lucheng district of Tongzhou had been relocated, their buildings demolished in preparation for developing the new municipal government offices.

Meanwhile, the population of Tongzhou New City has increased rapidly. From 2005 to 2013 the number of long-term inhabitants increased from 860,000 (22 percent nonregistered) to 1.36 million (nearly 40 percent nonregistered) (Peng 2014: 168). About half of these "migrants" to Tongzhou come from other provinces, and the other half from Beijing proper, using the new town either as a gateway to the capital or as an affordable second home or real

2.6. Tongzhou Canal in 2007 (photo by author). Author requests to the developer AECOM to publish its 2012 concept sketches of Tongzhou as a low-carbon metropolis were denied.

estate investment. As housing costs continue to escalate, however, such opportunities are increasingly limited to the wealthy and powerful. In fact, after the Beijing Municipality made public in 2015 its plans to move its government offices to Tongzhou, real estate prices skyrocketed even more. Tongzhou was originally planned to address social housing needs, yet recent studies indicate that many low-income housing projects are not being realized (Huang and Li 2014; Li and Chand 2013). As in Lingang, where "people" seemed absent from planning considerations, ordinary citizens may find themselves overlooked in government-formed alliances with corporate entities for the creation of megaprojects oriented toward regional and national competition rather than social and ecological sustainability. Indeed, scientific studies warn that the ongoing aggressive development of the region will cause Tongzhou district's ecological deficits to become even more severe, as urban land continues to expand at alarming rates, further eliminating arable land and fresh water sources (Wang et al. 2012; Wei and Ye 2014).

Dujiangyan New City and Chengdu CURD

With Beijing Municipality touted as the national model for sustainable new town development, and Shanghai Municipality for marine eco-city development, Chengdu Municipality is considered the national model for Garden City development. Chengdu first promoted urban-rural integration in 2003, and suburbanization intensified in 2007 when Chengdu and Chongqing established the first national experimental zones in rural land transfer, known as National Comprehensive CURD (Coordinated Urban-Rural Development) Experimental Zones. Dujiangyan, located forty-three miles from Chengdu, is a historic settlement due to its renowned irrigation system, which has diverted water from the Min River to irrigate the West Sichuan Plain for thousands of years.[7] Due to its UNESCO World Heritage Site status, its cultural legacy of hydraulics, and its unique agricultural practices and ecological resources, Dujiangyan plays a significant role in the region despite its relatively small population (680,000) and size.

The devastating Wenchuan earthquake on May 12, 2008, which destroyed 30 percent of Dujiangyan (figures 2.7, 2.8, and 2.9), further accelerated the city's regional integration with Chengdu. Less than a month after the quake, the State Council of the People's Republic of China approved a major ordinance initiating a centrally coordinated massive and rapid urban reconstruction of Chengdu Municipality. It governed all policies and schedules for earthquake reconstruction, including spatial layout, urban housing, urban construction, rural construction, public services, infrastructure, industrial recovery, disaster prevention and reduction, ecological environments, mental care, policy measures, reconstruction funds, and plan implementation (Guo 2012). Shanghai Municipality was appointed by the central government to assist with the first three years of the reconstruction, and the Chengdu Municipal Planning Department contracted Tongji University to reconstruct Dujiangyan. According to a timeline presented by Professor Dai Shenzhi of the Tongji University College of Architecture and Planning, who oversaw the reconstruction of Dujiangyan, his team of planners arrived a mere five days after the earthquake. Their authority to plan was officially granted by the State Council Ordinance on June 7, 2008, and from June 9–23 the team conducted surveys and investigations and completed a draft plan (Dai 2012). On July 17, 2008, the formal redevelopment plan for Dujiangyan was presented in Chengdu to international scholars, including myself, who were attending the Tsinghua-MIT China Planning Network conference.

In addition to the sheer speed of its completion (two months after the quake), the July 2008 Dujiangyan plan was striking for its similarity to new

2.7, 2.8, and 2.9. Thirty percent of Dujiangyan was destroyed by the May 12, 2008, Wenchuan earthquake. Photos by Robert Olshansky (July 18, 2008).

2.10. Postdisaster reconstruction scenario for the Dujiangyan Ecological Garden City waterfront. Source: Chengdu Planning and Management Bureau, 2008 (Guo 2012: 48).

town development plans in Shanghai and Beijing at that time. The plan rationale indicated the city must expand from thirty square kilometers to sixty square kilometers by 2020 in order to accommodate five hundred thousand additional residents. In the July 2008 plan, the city's original residents were to be located to a new district (Dujiangyan New City), while Dujiangyan Old Town (later known as Dujiangyan Ecological Garden City) was to be redeveloped as an "international tourist center and historical and cultural city" (figure 2.10). In the plan's rhetoric, "pursuant to a 'highlighting mountain and river' principle, administrative, industrial, and secondary school functions will be removed."[8] In other words, the old town would be gentrified, with its functions primarily designated for tourist consumerism and, like Beijing before the 2008 Olympics, it would relocate its urban poor to an autonomous new town. The municipality's postdisaster strategy promoted the area's ecological and cultural resources in order to attract tourism and speed the economic recovery of the region. It initially aimed to attract domestic tourists emotionally motivated to contribute to the postearthquake recovery by visiting the disaster area and understanding the local situation through personal experience. Dujiangyan was ultimately to be branded as an international tourist eco-city while attracting local day-trippers from Chengdu. On July 17, 2008, the Chengdu Transportation Department presented our visiting delegation with its already completed plans for the Chengdu-Dujiangyan express railway, equipped for speeds up to two hundred kilometers per hour.

Such aspirations notwithstanding, as of 2014 the attempts to recast Dujiangyan as a leading eco-tourist destination appear to face many challenges. A scientific study of the resulting Ecosystem Services Value (ESV) change due to the land use planning of disaster reconstruction (He 2012) and a sociological study of ecological damage caused by tourists and residents (Wang 2012) are both cautionary in tone. Another article warns that hotel stock in New Dujiangyan City is grossly underutilized because most tourists are day-trippers from Chengdu (Deng 2012). Yet another study denounces culture shows in Dujiangyan, which feature its unique "water" and "Taoist" cultural elements, as "stereotyped" and "boring." The leading show, "Detailed Description of Dujiangyan" (modeled on Zhang Yimou's wildly successful *Impression* culture shows such as his *Impression Liu Sanjie* [*Yinxiang Liu sanjie*] in Yangshuo, Guilin), was singled out for the harshest critique (Guo et al. 2012).

Furthermore, the new city designs utterly fail to account for the local housing style, regional context, and county-level city scale (Yan 2012). The unique rural landscape of the Chengdu plain comprises the *linpan*, a special layout consisting of individual farmhouses surrounded by wooded areas, the irrigation zanja, and the fertile farmland area dotted by scattered villages. However, with the transfer of rural land to Chengdu Municipality (as part of CURD), the urban development along the transportation corridor of the high-speed railway between Chengdu and Dujiangyan is radically altering cultural heritage. Yang Qingjuan and coauthors describe the original rural morphology as "patches (the vast plain of farm land), corridors (roads and irrigation canals), and matrices (small matrices are thousands of *linpan*, big ones are towns)" (2011: 782). These scholars decry the destruction of the *linpan*, nourished by a hierarchical irrigation system of canals and zanjas from the Dujiangyan irrigation system that cover the plain like a "circulatory system."

Others criticize the redevelopment of Dujiangyan because of its forced relocations and violation of residents' property rights (Deng 2010; Flower 2009).

Interestingly, Professor Dai Shenzhi of Tongji University (2012) acknowledged that results of surveys conducted by his Tongji University planning team showed that 93 percent of Dujiangyan's original residents refused to relocate. Yet the redevelopment plans for the old town had clearly already been established, with the surveys a mere formality. Indeed, less than one month after the earthquake the American Chamber of Commerce in Shanghai was already promoting the following investment opportunities in the Dujiangyan locale:

Star-ranked luxury hotels/commercial projects built on land originally occupied by:

1. People's Hospital, Beijie Primary School and Dujiangyan Machine Tool Factory
2. Bojiang riverbank in Anlong Town, Juyuan new zone, Cuiyuehu Town
3. Lianhua Lake in Xiang'e Town, Qingcheng Lake in Liujie Town
4. Qingchengshan Town, Qingchengshan Software Industrial Park

Usable construction land:

1. Qingchengshan Software Industrial Park Hotel: state-owned land of 4 acres
2. Ziqidonglai Holiday Hotel: state-owned land of 3.8 acres
3. People's Hospital: state-owned land of 3.4 acres
4. Beijie Primary School: state-owned land of 2.9 acres
5. Dujiangyan Machine Factory: state-owned land of 3.3 acres.

Despite the genuine outpouring of goodwill and concern by domestic tourists who came to provide aid in the aftermath of the earthquake, there is no question the disaster was immediately seen as an opportunity for Chengdu Municipality to transfer rural land into developable urban land in and around Dujiangyan. Dujiangyan planners provided extensive economic, cultural, and geological rationales for its plan to double in size by 2020. Nonetheless, the June 2008 list of real estate development opportunities on the American Chamber of Commerce website tells a primarily speculative, less scientifically driven, story.

Building for What and Whom?

New city development in the twenty-first century was revived as "eco-city" fever in the wake of a central government initiative promoting urban-rural integration as a strategy to regain control over the governance of rural land. In

2002 a "province-manage-county" (PMC) reform was initiated in select provinces and was later expanded during the Eleventh Five-Year Plan (2006–2011) to include eighteen provinces and province-level cities. The PMC policy was instituted, in part, to reverse the radical devolution of power to local municipalities that had been formalized by an unprecedented structural change in administration in 1983, when the "cities leading counties" policy "ushered in an epoch of city-led regional development for the first time in Chinese history" (Zhu 1999: 30).

The 2002 PMC policy was an attempt, then, to regain control of the development of rural land. Publicly it promotes "coordinated urban-rural development" as a means of redressing rapidly rising income disparities (by 2009 the per capita urban-rural income ratio had risen to 3.33). Its wide-ranging policies address public finance, local public administration, social systems, rural land use, physical development, and changes to the *hukou* (household registration) system. In 2007, National Comprehensive CURD Experimental Zones were established in Chongqing and Chengdu, and the experimental zones were soon to be replicated elsewhere. Yet, according to a recent study by Nick Smith, Chongqing's reform program has been faltering. Smith reports that during the first sixteen months more than 3.2 million people traded in their rural registration to become officially urban, but since then the number of applications has dropped precipitously, to just four hundred thousand in 2012. A 2010 survey conducted by the municipality's Public Security Bureau found that 90 percent of Chongqing's rural residents were unwilling to participate in the program. As Smith argues, "these numbers call into question Chongqing's ambitious 2020 goal and, by extension, the success of China's national urbanization project" (2014, 370). Smith elaborates further:

> By flexibly moving between urban and rural institutions and locales, residents challenge norms commonly articulated through discourses of "quality" (*suzhi*), a term that indexes social and cultural difference within Chinese society. These new hybrid identities thus destabilized the very urban–rural distinctions that produced their marginalization in the first place. (2014, 377)

Thus at the level of realpolitik one questions the motivations for China's new city development in particular, and its urban-rural integration policies in general. As noted previously, Chinese new city development in the twenty-first century extends planning and policy rhetorical legacies from the early socialist period. Initially new urban sectors were developed within, or adjacent to, existing cities or towns in order to accommodate the Stalinist-style architecture associated with the key bureaucratic institutions of the Party. Furthermore, the "new cities" established in frontier areas (such as Dayanzhen, prefectural seat for the Lijiang basin in Yunnan) became coded as Han Chinese–identified areas given the in-migration of outsider technocrats and military personnel (White 2010: 151). The strategy of resolving the "nationality problem" of a multicultural socialist nation-state by bringing socialist urban modernity into the lives of ethnic minorities replicates the logic of bringing socialist urban modernity to rural peasants, be they Han or otherwise. While Hu Jintao's earlier policies ostensibly emphasized rural development of the "new socialist countryside," many considered his subsequent emphasis on "urban-rural integration" to be the "most explicit affirmation yet of the view that urban markets and urban administration are the keys to escaping poverty" (Abramson and Qi 2011: 507). Thus postearthquake reconstruction and development of impoverished Western regions often proceed via a policy of "pair assistance," where a wealthier Eastern municipality invests 1 percent of its GDP in a developing (Western) region.

The downside to such policies is that new towns are often "mechanically reproduced" by planning departments in Eastern academies such as Tongji University or Tsinghua University, with little regard to local scale, geographic, or historical conditions. Local residents are often forced to relocate or are incentivized to convert rural *hukou* to urban ones so that municipal officials can develop land previously owned by villages, or relocate the urban poor (and their polluting industries) to the outskirts of increasingly gentrified city centers. Further, this becomes a way to incorporate the "informal" (rural migrant) citizenry under state control, and it opens up new markets for services such as privatized health insurance. To rationalize these actions, municipal government officials appeal to familiar rhetoric about the "chaos" caused by "informality" and the need to raise the "quality" of the body politic. If the farmer-peasants were to educate the intellectuals and artists how to be good socialists during the Cultural Revolution, the urban bourgeoisie are now tasked with raising the "quality" of the farmers.

In keeping with this logic, twenty-first-century suburbanization strategies in China are manifesting unprecedented forms of virtual urbanism predicated upon consumption of the virtual rural. In the most extreme cases, laboring bodies produce "new cities" and "virtual countrysides" without inhabiting them bodily. Thus the question remains—Why is so much building going on, and for whom? When I put the question of "who will inhabit Lingang" to my undergraduates in 2013, one from Taiwan said, "My uncle regrets sinking his money in Lingang real estate," while a student from China retorted, "Oh, he'll get a return on his investment, but it will not be for another thirty years when people actually want to live there." Are cities such as Thames Town and Lingang merely Potemkinism? Do utopian cities such as Lingang, Tongzhou, and Dujiangyan simply provide a rationale for China's strategy to urbanize vast metropolitan regions by converting rural land for urban development along major transportation corridors to a major metropolis such as Shanghai, Chengdu, and Beijing?

Thirty years from now the land between Lingang and Shanghai, between Tongzhou and Beijing, and between Dujiangyan and Chengdu will no doubt be fully developed. But the social vitality of these new urban developments, and the long-term sustainability of these planned "eco-cities" and "ecological regions" is unlikely. A recent review of the scientific literature on China's environmental conditions concludes that "the ecological footprints of China and most provinces run a national or regional ecological deficit" (Wei and Ye 2014: 761). Most studies of Chinese eco-cities (Chien 2013; Kim 2014) are equally sobering. Kim, for example, concludes that the rhetoric of green city development ignores those whose livelihood depends on sustainable social, economic, and environmental conditions, instead codifying "green" into modernist design, scientific abstraction, and efficiency-oriented managerialism: "The current discourse of 'sustainability' in these cities therefore fails to distance itself from the developmental paradigm of past decades, when much of Northeast Asia's rapid urbanization, as the quintessential program for nation building, foisted unresolved socio-economic distress on its citizens" (Kim 2013: 350).

In sum, eco-city development, sustainability discourse, and policies of urban-rural integration and suburban land development increasingly benefit the few at the expense of the many. These policies are maintained via legacies of neoliberal pseudo-scientific thought that rationalizes social engineering in Chinese society and beyond. Attempts to trade in the natural environment and human societies as commodities are intensifying as global interests incentivize virtual performances over embodied productions. While popular resistance to elite attempts to normalize the society of the spectacle arises continually within China and elsewhere these efforts have not yet fundamentally altered carbon emissions or social inequities. Instead, global ecological conditions continue to deteriorate at alarming rates as planetary urbanization intensifies.

Notes

1 As of 2007 more than half of the world's population lives in cities; China surpassed 50 percent urbanization in 2010.
2 See, e.g., Keil (2013).
3 Each "eco-city" has recently been branded as such (hence the names in parentheses), but for consistency I will refer to the three cities as "Lingang New City," "Tongzhou New City," and "Dujiangyan New City." Further, Lingang was renamed Nanhui in 2012, but it is still popularly referred to as Lingang.
4 Press release (May 2, 2004) posted on the website of Burchill VDM after it was awarded the contract with Tongji University for building Yangshan New City. Copies on file with author.
5 The Tongzhou Master Plan (2012) is posted on the website of Laguarda.Low Architects LLC: www.laguardalow.com/projects/asia/tongzhou-master-plan/.
6 See the AECOM website: http://www.aecom.com/What%20We%20Do/Design%20and%20Planning/_news/Building%20a%20low%20carbon%20metropolis. Copies on file with the author.
7 The Dujiangyan irrigation system, the first hydraulic irrigation system in human history, was designated a UNESCO World Heritage Site in 2000.
8 Quoted from the July 17, 2008, Dujiangyan Reconstruction Plan, presented by Chengdu Municipality (in the author's personal files). See also the rationale for tourism by the Chengdu-based academics Yan, Wang, and Chen (2011).

3

HEGEL'S PORTFOLIO

Real Estate and Consciousness

in Contemporary Shanghai

Alexander Des Forges

To read and understand *New Shanghai People* is to read and understand the transformation and progress of our time.
—*New Shanghai People* (back cover)

Meili xin shijie (Beautiful new world; dir. Shi Runjiu, 1999) opens with Zhang Baogen on a bus from the countryside to Shanghai. Baogen has won a lottery for a luxury apartment in the city, but he finds to his great surprise when he shows up to claim his award that his apartment is an "apartment yet to be built" (*qifang*). Although much fun is had with Baogen's initial inability to comprehend this use of the term *qi* to modify *fang*, he is shrewd enough to resist the developer's offers to buy him out. Instead of accepting a cash payment for much less than the supposed market value of the apartment, Baogen settles in with a distant relative who is a Shanghai native, his "little auntie" Jinfang, to wait the developer out. This decision provides viewers with a series of fish-out-of-water vignettes, as Baogen slowly learns how to function in his new setting, but the film also chronicles Baogen's own transformative effect on Jinfang, resulting in a highly predictable romantic denouement. In the process, the film raises interesting questions about the future of Shanghai identity in a time of renewed in-migration to the city, questions that are explored from a different perspective in *Xin Shanghairen* (New Shanghai people, 2003), a volume of essays published four years later.

Beautiful New World appeared at the end of a decade in which discourse on Shanghai identity flourished at multiple levels. Although the city kept a relatively low profile in the 1980s as the Pearl River Delta surged economically, *gangtai* (Hong Kong/Taiwan) music and film oriented popular culture toward Hong Kong and Taiwan, and the *xungen* ("roots-seeking") movement suggested that essential cultural truths were to be found in the interior, not along the seaboard. The opening of Pudong to large-scale development in 1990, however, brought national and international attention back to the city, and loosened restrictions on in-migration returned the question of "Shanghai identity" to a prominent place within national and local cultural production. This question runs through a wide range of works in the 1990s, from Wang Anyi's *Youshang de niandai* (The age of melancholy), *Xunzhao Shanghai* (In search of Shanghai), and *Changhen ge* (Song of everlasting sorrow); to the hit dramatic television series *Niezhai* (Sinful debt, 1995), which centered on stories of children of Shanghai rusticated youth coming to Shanghai to seek out their parents; and from books claiming to reveal the secrets of the Shanghai personality to radio shows using Shanghai dialect. *Beautiful New World* builds on this trend, and adds its own distinctive answers to the question of Shanghai identity through its organization around a double pedagogical project, as each of the principals has something important to learn from the other.

It would be a mistake, however, to conclude that since Baogen and Jinfang each have something to teach the other, the two curricula are therefore of equal value. Instead, by the end of the film it is clear that the petty skills that Baogen learns from Jinfang are much less significant than the larger life lessons that he has to offer her. This inversion of a classic trope—the education of a rural newcomer to Shanghai—clearly finds its moment at the turn of the twenty-first century, as a valorization of newcomers to the city also forms the basic premise on which the collection *New Shanghai People* is organized. Unlike blander technocratic works such as Wang Guorong's *Kuashiji de Shanghai dushi wenhua xingxiang* (1998; Shanghai's turn-of-the-century urban cultural form) that seek to understand and manage the city through numerical data and imagine its future development in the form of airports and high-speed Internet access, *New Shanghai People* is unabashedly philosophical in emphasis—over half of the essays are by scholars housed in departments of philosophy, literature, and history, including the noted historian Xiong Yuezhi and the author Cheng Naishan. Particularly striking is the prominent representation of participants in the *renwen jingshen* (humanist spirit) debates that originated in

Shanghai just under a decade previously, including Chen Sihe, Zhang Rulun, and Zhu Xueqin. In its attempt to specify modes of cultural production and consciousness appropriate for nonstate actors, this volume revisits some of the complicated questions about the role of the market characteristic of the earlier humanist spirit debates, but at a time when the social role of capital is significantly less subject to challenge.[1]

Shanghai's rise in the late Qing and the early Republican era and its dramatic "return" in the 1990s and early 2000s have often been understood as direct expressions of tendencies in the global market. By extension, its relative eclipse during the second half of the twentieth century must then be explained as a distortion or willful violation of these natural tendencies, whether that distortion takes the form of a strategic relocation of factories to cities in the interior in the 1950s, the restriction of in-migration to the city beginning in 1958, the rustication of educated youth in the 1960s and early 1970s, or state emphasis on development in the Pearl River Delta during the 1980s (Des Forges 2007: 180–183; Huang 2004: 99–118; Xiong 2003). But what if we were to think of willed, conscious action as integral to the city's return and future development as well: not just in terms of the aims and tactics of individuals or plans drawn up by government agencies, but as a broader purpose that finds its expression in and through the city's development? Such a metanarrative is to be found throughout the majority of essays collected in *New Shanghai People*: Paradoxical constructions of the relations between conscious will and its material circumstances that are reminiscent of Hegel's account of the interaction between the Spirit and the Substantial ground appeals to Shanghai people to exert themselves to respond to the demands of changing national and global economic conditions.

Shanghai People between the City and the Nation

Without exception, the essays in *New Shanghai People* strive to tie the city firmly to the nation, staging this relationship in most cases as a specific form of exchange: Shanghai is continuously renewed through in-migration from elsewhere in China; in turn, the city has a special responsibility, duty, or even mission, to repay (*jiaodai*) the nation and History by taking a leadership role in economic and cultural development. Although the term *xin Shanghairen* is defined variously by the different authors, it is clear that this understanding of the relationship between the city and the nation leads to a narrow focus on one subgroup of the city's population: white-collar workers and entrepreneurs who have come to the city from other regions in China since 1990.

In striking contrast, natives of the city—which is to say, "Old Shanghai People"—play a less than glorious role: The generations that lived through the 1950s, 1960s, and 1970s in particular are stereotyped as prejudiced, small-minded, unimaginative, and untrustworthy. Whether they appear as straw figures against which the virtues of in-migrants to the city are set off, as passive partners in the enterprise of creating a new city, or, in the case of the liveliest members of the younger generations, as potential full participants in the city's new development as long as they work to conquer their negative qualities, those born in the city are distinctly secondary to the new arrivals, in part because they *happen* to live in Shanghai rather than having actively *chosen* to live there. Dai Zhikang goes so far as to argue in the introduction to the *New Shanghai People* volume that it is precisely through the choice to relocate to Shanghai that in-migrants from other regions of China may well demonstrate a deeper and more thorough understanding of the city's potential than those who are merely born and raised there (2003b: 1–2).

This deeper conceptual grasp of the city's essence—whether implicit or explicit—is contrasted throughout the volume with the petty material tactics and practices of Old Shanghai People, whose identity is thought to be based as much on unreasoned discrimination against newcomers who have not mastered the peculiarities of the local practices, foremost among them Shanghai dialect, as it is on any set of skills that will be of use in the city's continued development. Shanghai dialect figures prominently in modern Chinese cultural production generally as concrete evidence of the distance between Shanghai natives and the rest of the nation, with one side quick to use it as an in-group language, and the other side unwilling to hear it spoken in person or on screen, even when subtitles are provided. As an indicator of Shanghai identity, its efficacy is unparalleled, and its place in the process of national cultural production is consequently never uncontested. The television series *Sinful Debt*, with a narrative predicated on the radical disjunction between Shanghai and Xishuangbanna, was originally filmed in Shanghai dialect for regional broadcast, but when it was rebroadcast in other cities the dialect was replaced by standard Mandarin (Ye Xin 2001). And although *Beautiful New World* makes extensive reference to standard Shanghai tropes, from Jinfang's habit of wearing pajamas around the distinctive *lilong* neighborhood to the fetishization of Western liquor and cigarettes to the use of a "Suzhou-style" teahouse performance as a framing device, Shanghai dialect is virtually absent from the film. Even conversations between longtime residents of Jinfang's *lilong* are all conducted in standard Mandarin. *New Shanghai People* continues this generally dismissive approach to Shanghai dialect: Although one of the central questions that would seem to be raised by the massive in-migration to the city is whether Shanghai dialect will continue to function as a vital part of the city's identity, the authors show no interest in this topic, mentioning local dialect only in passing either as a literal means for discrimination against outsiders (Cao 2003: 102) or as a figure for prejudiced attitudes in general. This treatment of dialect provides us with an important hint of a broad tendency in discourse on New Shanghai People: Dialect speech is dismissed because it helps to constitute the old Shanghai identity that is to be overcome; but even more so, it is neglected, as it represents cultural materiality itself as a category that these authors take little interest in.

Old Shanghai People not only maintain a problematic fixation on a narrow conception of the region but, according to Chen Xueming, also have the wrong attitude for a new global era. Their "slave quality" (*nuxing*) can be seen even in their real estate preferences: apartment complexes that claim to have some connection with the West, no matter how tenuous (2003: 10–11). By simultaneously privileging their own region over other places in China on one hand, and elevating the West over China in the international sphere on the other, Shanghai people seem to be guilty of a particular disregard for the nation. Although no other essay puts it as sharply as Chen Xueming's does, it is clear that the question of Shanghai's place within China is directly relevant to much of the discussion of what a new Shanghai identity could be.

Tsung-yi Michelle Huang has written that, since the 1990s, Shanghai's urban development has been "predominantly modeled on the image of a generic global city" in an attempt to mirror cities such as New York, London, Tokyo, or Hong Kong (2004: 100–101, 103–107). Contributors to *New Shanghai People* manifest their particular tension over a perceived disjunction between what is and what could be by arguing that if this is indeed the case, it should not remain so, or Shanghai will not realize its full potential:

> If Shanghai people can really in a relatively short period of time raise up a city that can be called a second New York on the west coast of the Pacific, then this generation of Shanghai people can all in all be said to have given history what was minimally required (*zongsuan youle yige jiaodai*). But if Shanghai people want to truly pay history and Chinese civilization back ... [they should] instead of simply copying a Western model develop Shanghai into a metropolis that has special characteristics of the Chinese nation and can rival Western cities. (Chen Xueming 2003: 3)

Where city planners see Pudong's similarity to other global cities as key to its role as the "dragon head" leading change in the Yangtze River Delta (and, by extension, the nation as a whole), *New Shanghai People* authors see the city as maintaining its rightful position in the vanguard only by going beyond mere mimicry and establishing its own distinctive brand of cultural production in the broader global context (Wang 2003: 22–23; Zhang 2003: 61). What is at stake here is clear from Zhang Rulun's assertion: "Shanghai is China's Shanghai" ("*Shanghai shi Zhongguo de Shanghai*"); while it is possible to imagine a similar comment being made about Hong Kong, it is harder to imagine its relevance in the case of Beijing, Chongqing, or Changsha. Zhang's concern for Shanghai's national pedigree extends to his attention to the historical record when he reminds us—quite accurately—that the "adventurers" for whom Shanghai was a "playground" hailed not only from foreign shores, but also, in greater numbers and to greater effect, from elsewhere in China (58–59, 61).

A contentious relationship between "China" and "the West" is explicitly invoked throughout *New Shanghai People*. Left implicit is a similar dynamic in which the economic and cultural forces of overseas Chinese and Taiwanese "compatriots," that part of Chinese transnationalism and the Sinophone sphere that lies outside the boundaries of the People's Republic, contest the very definition of "China" (Ong 1997: 175). Although there are scattered remarks dismissing Hong Kong and Singapore as models for Shanghai's future development, there is almost no discussion of the specifically transnational aspects of Shanghai subjectivity that Mayfair Yang identifies beginning in the mid-1990s (Yang 1997), not to mention the cross-straits complexities arising from Taiwanese capital and entrepreneurs active in the city. The terms of the encounter are resolutely bilateral, with in-migrants to Shanghai playing a crucial role in balancing the powerful influence of "the West." As Chen Sihe puts it in his essay for the *New Shanghai People* volume,

> If we really want to further Shanghai's cultural construction, especially the strategic development of Shanghai culture after entrance into the WTO, in addition to relying on state leadership, we must consider how best to unleash spontaneous cultural forces among the people (*minjian*) to create a lively and diverse cultural environment, and have the masses (*minzhong*) consciously develop national and regional cultural traditions; this will have real significance in counterbalancing the floodtide of Western culture pouring into the marketplace. (2003: 94)

In this way, the discourse of New Shanghai People negotiates an accommodation with nationalist discourse that may be suspicious of the city for its semicolonial history by integrating and repurposing it, rather than challenging it directly. As was the case in the late Qing and early Republican period, Shanghai is posed as an arena for productive confrontation, in which centers of power that are presumed to operate in opposition are forced into an uneasy balance, which will yet turn out to be productive for the city, if not so much for other regions (Des Forges 2007: 29–55, 114–121).

Who Is Left Out?

Do the "masses" (*minzhong*) that Chen Sihe sees as the counterweight to Western culture in Shanghai include migrant workers (*mingong*)? Of the new arrivals to the city in the 1990s and 2000s, the majority were migrant workers, but their position in the new Shanghai was quite complicated: It is fair to say at the very least that local governments make substantially less effort to accommodate their needs than they do those of newly arrived white-collar employees. As Xiong Yuezhi demonstrates, steps to ease residence restrictions in order to encourage economic development are taken first for individuals with a college education or capital to invest, and only later (if at all) for ordinary workers (2003: 119–120). In the same vein, Michelle Huang quotes an official saying in 2001 that the aim is "to replace 4 million migrant workers . . . with 4 million professionals" (2004: 112). The tension and contradictions characteristic of official policy and social attitudes manifest themselves as well in *New Shanghai People*, few of whose authors directly address the question of where *mingong* might find a place in the figure of the New Shanghai Person.

Wang Defeng praises the "life force" of new arrivals to the city, dividing them into "three high-level" categories (*san gao*): namely, individuals with high educational attainment, substantial amounts of capital, or high levels of physical strength (2003: 22–23). Curiously, however, the third category is then explained as "blue collar workers in the various enterprises and small tradesmen in business for themselves," which leaves unresolved the question of whether or not a major category of *mingong* such as construction workers is included. Deng Weizhi and Bao Zonghao are more exclusive: Deng finds Internet usage as a mode of production to be a defining characteristic of New Shanghai People, while Bao, as we will see in more detail later, focuses narrowly on the executive and managerial classes as "representative" of the city's new identity; *mingong* do not merit a mention in either essay. Most telling is a brief passage in Chen Sihe's "Shanghai People, Shanghai Culture, and Shanghai Intellectuals": "When we spit out a disdainful '*waidi ren*' ('outsiders') on seeing *mingong* and *mangliu* from other regions on the street or on the bus, we seem to have forgotten our own résumés. . . . No one can deny that after some years the children of these *waidi ren* will reside in Shanghai, go to school, grow up, and become a new generation of Shanghai people" (82).

Of the various mentions of *mingong* in the volume, this passage best captures an underlying attitude that appears in diverse manifestations: Physical laborers are integral to the New Shanghai project—to deny this fact is impossible—yet they appear invariably at a remove. Like the Old Shanghai Person, the migrant laborer cannot be considered a full member of the new generation, though his or her children will most likely be, an assumption of the sort addressed in more detail in Carlos Rojas's and Ralph Litzinger's chapters in this volume. But this trope for the most part elides the difficulties involved in this process: How will those sons and daughters eventually gain Shanghai residence permits? And which schools will they attend in the meantime?

Given the emphasis on interaction and exchange between Shanghai and other regions in China central to any definition of New Shanghai People, the lack of attention to returned youth and their children is also striking. High-school and college-age individuals who "went to the countryside" in the late 1960s and early 1970s could certainly be thought of as analogous examples of the combination of "Shanghai" and "outsider" traits; their experiences on return, and those of their children, have been a frequent topic in Shanghai fiction and essays. Although staged in dramatic terms exaggerated for effect— as suggested by its title—*Sinful Debt* reminds us that some of the very same questions about the experiences and potential contributions of *waidi ren* to the contemporary city are raised by these returnees. But neither the experiences of returned youth in the 1980s and 1990s, which could prefigure the struggles of later arrivals from other regions to locate housing, nor the experiences of their children, for whom Shanghai can be counted as a "native place" yet becomes their actual place of residence only after they attain adulthood, figure substantially in the discourse of New Shanghai People. Returnees and their children call into question the clarity with which an Old Shanghai identity can be distinguished from a New Shanghai identity that is in the process of formation subsequent to the supposed transformation of a planned economy into a market economy.

In addition, as mentioned previously, contemporary in-migrants are characterized on an individual basis as having actively chosen to relocate to Shanghai;

at the same time, when they are considered as a group, this choice is seen to reflect "natural" tendencies of the Reform Era national market. Under these terms, returned youth are doubly excluded from discussion: Their original departure from the city is understood to have been involuntary and political in nature rather than a "free" choice under economic law, while their return is likewise not considered a deliberate choice between reasonable alternatives for future development but rather an outcome to be expected from their attachment to their family, friends, and the city *as it was*. If in-migrants owe their allegiance to the city's future in the abstract, as suggested by the opinions expressed in national surveys on Shanghai's "livability," in many cases by people who have not yet visited the city, returnees, fairly or not, most likely are thought to orient themselves to concrete remnants of the past, whether in the aesthetics of daily life or the use of Shanghai dialect to interact socially.[2] The difference is between the city as concept and the city as lived experience and memory.

Urban Development as a Revolution in Consciousness

Most essays in *New Shanghai People* frame Shanghai's future in two apparently contradictory modes: one necessary and the other conditional.[3] First, the authors are quick to argue that Shanghai's current rapid material development is entirely natural and that their interest in the city stems from this "objective reality" rather than from subjective prejudice. As Wang Defeng puts it in "'New Shanghai People' and Contemporary Chinese Cultural Life":

> The new Reform Era has created a new inmigrant movement to Shanghai; this is not, in fact, the result of subjective effort on anyone's part. In its function as the will of the nation, national policy is merely consciously responding to (*shunying*) an objective social movement. This is the historical materialist perspective. From this perspective we can see the consistency of this movement with the basic characteristics of the contemporary development of the market economy. (23)

But the proposition that Shanghai must become the dominant metropolis in China and a key player on the world stage is advanced first as a given only to be called into question through a reflection on the insufficiencies of existing Shanghai people. The pivot point between these two modes often comes down to expressions like *zhiyou* (only when . . .) and *caineng* (will it then be possible . . .): "That Shanghai will in the future attain the status of a 'world city' cannot be doubted. But in order to become a world city, hardware alone is insufficient, there must be software to support it. Only when there is software will it then be possible for the reality of Shanghai to match its name as an international metropolis" (Zhang 2003: 57). Putting a twist on categories familiar from dialectical materialism, Zhang informs us that "internal quality" has not kept pace with the improvements in "external circumstances" (60); Chen Sihe likewise suggests that the connection between development of "material culture" and "spiritual culture" is not automatic (86); even Wang Defeng ends with an exhortation to build a new culture for the city so as to avoid a "historical tragedy" (24).

Caught between the historical necessity that Shanghai become a leading global city, and the perceived shortcomings of Shanghai people in the face of this necessity, the authors focus as much on what Shanghai people should become as on what they in fact are. Hu Shouyun, for example, dismisses the question of how to define "Shanghai people" as a group as needlessly complicated, and turns instead to the question of what consciousness an individual needs to have in order to live a (productive) life in the city in the early twenty-first century (2003: 39–41). Compared to the complexity of trying to adequately characterize the diverse population, articulating what this population needs to do next to meet the tasks that history sets it seems both simpler and more useful.

Written at a moment when the state still maintains substantial capabilities to regulate the behavior of individuals and markets, these essays are striking in their relative lack of interest in concrete measures that state actors might take to encourage Shanghai's social, economic, and cultural development. With the exception of Hua Min and Dai Zhikang's calls for more state investment in infrastructure (Dai 2003a: 210; Hua 2003: 201), the authors direct their recommendations not at officials, but rather at Shanghai residents and potential residents en masse whose moment-by-moment interactions "create" the city: "Whatever kind of people you have, you will have that kind of Shanghai. If you say that the Shanghai of the 1980s and 1990s is a 'material transformation' of 'Shanghai people's' life and efforts of that time, then in a new century in which the forms of living and production have fundamentally changed, 'Shanghai people' must consciously change and renew themselves as 'New Shanghai People'!" (Gu 2003: 55).

Indeed, one gets the distinct sense that actions taken by the state constitute not so much expressions of will in any sense, but merely one more dimension of the material circumstances under which that will appears and becomes conscious of itself (here there are interesting echoes of Hegel's critique of the Chinese state as mere substance, lacking reflexivity and true agency). Agency reposes in the people of Shanghai themselves; as a result, they are to be tasked with creating their own new consciousness.

In other words, the development of the New Shanghai Person is a willed project rather than a "natural" process—as such it recalls the movement paradigm articulated in the 1940s and 1950s, in which the broad masses are mobilized and inspired to raise their consciousness to a higher and more productive level. Yet unlike the movement tradition, in which careful study of prescribed texts and repeated articulations of one's own history orally and in writing serve as the concrete practice grounding the wished-for transformation in consciousness, the forging of a new Shanghai consciousness is imagined to take place primarily in the smithy of the mind itself, a willed leap from false consciousness to true that comes from no recommended practices. Zhang Rulun, for example, finds the consciousness expressed (often implicitly) in the views of ordinary Shanghai residents to be a central concern: "It is these views (*kanfa*) of Shanghai residents, rather than their words and actions, that decide the city's appearance and its fate" (2003: 58). He is careful to trace the narrow worldview of the average Shanghai resident to excessive exposure to broadcast media, in particular, television shows that construct an unrepresentative image of the world outside; this is a problem that must be dealt with. But the solution lacks this specificity: Shanghai residents are not told to turn off the television, visit the library, or seek pen pals outside of cities like New York and Los Angeles, but rather merely to *realize* that there is more to the world. Similarly, Yang Guorong writes in "Shanghai People: History and Today" that New Shanghai People must have global vision (*yanguang*) and consciousness (*yishi*), which in turn call for an open and tolerant mind (*xionghuai*) (2003: 127). In effect, as we see most clearly in Chen Sihe's essay, the concrete practice of everyday life is subordinated to the abstract goal of development; Chen reinscribes the split between material circumstances and the response of the individual or collective will to those circumstances in the form of cultural development *within* that cultural development itself—only that part of culture furthest removed from the material maintains critical and constructive potential.

Despite the rhetoric of natural tendencies of market forces and the immutability of historical heritage, what many of these authors are in fact proposing is a Shanghai identity that is on the order of a collective performance, an *enactment* of Shanghai uniqueness and superiority that will live up to a preconceived idea of the city. In this collective performance, mobilizing state, market, and cultural forces, are the Shanghai residents that Gu Xiaoming refers to memorably as "vehicles of culture" (*wenhua zaiti*) best understood as individual performers or as mere instances of the transcendent spirit of the city's development? Zhang Rulun notes that Shanghai residents' worldview must be improved in order to construct a world city whose "name fits its reality" (*ming fu qi shi*). But as Zhang makes clear, the goal is not to fit the name to the reality, but rather to fit the reality to the name, to give the city the right to speak up and lay claim to its own name: "Shanghai must dare to confidently say, 'Shanghai *is* Shanghai' ('*Shanghai jiushi Shanghai*')" (2003: 60–61).

Where *New Shanghai People* remains abstract and conceptual, *Beautiful New World* is concrete: The film embodies the contest between Old and New Shanghai identities in the everyday details of interactions between Jinfang and Baogen. Rather than exhorting Jinfang to broaden her worldview, Baogen leads her to a new outlook by an accumulation of small examples: of his commitment to hard work in the box-lunch business he has created; of his basic loyalty, fairness, and optimistic take on human nature; and of his thorough investment in a grand vision that requires patience and sacrifices along the way. And although the gender politics are quite similar—Baogen finds inspiration at a crucial moment from a street musician who challenges him with the words "Are you a man?" even as Chen Sihe writes that the Shanghai people living in the city since the 1950s are better understood as the "little daughters-in-law" of the city rather than true *Shanghairen* in their failure to imagine appropriate directions for future development—the movie has a very different take on class distinctions (2003: 91). In *New Shanghai People*, there is a clear hierarchy among the "three high-level" categories: Workers who make use of their physical strength are ranked as substantially less important to the future of the city than those who have higher levels of education or significant accumulations of capital; depending on the context, *mingong* find themselves either in this lowest category or left out of the hierarchy of New Shanghai People entirely. *Beautiful New World*, by contrast, gives us a protagonist who is in at least some sense representative of most of the different groups who have a stake in Shanghai's future development, and his class identity therefore becomes much harder to pin down: Baogen is an entrepreneur, establishing a *minying* business; he also works the kinds of jobs that *mingong* often must; he aspires to future status as a "successful individual" (*chenggong renshi*) but has not yet attained it by the end of the film;[4] and as an individual who stubbornly refuses to give up his apartment in exchange for an unreasonably low cash offer, he resembles the *dingzi hu* (literally, "nail households") who became central to the Old Shanghai image by the first decade of the twenty-first century (Shao 2013). In this way, Baogen scrambles the categories into which the Shanghai urban experience is usually organized. As a "nail household," stubbornly clinging to what he has, he should be understood as an Old Shanghai Person, but he is not. As an outsider with a grand vision, he should fall under the rubric of the New Shanghai Person championed in *New Shanghai People*, but he is not "high-level" in educational attainment, and as Jinfang points out, his box-lunch business will never compel her to call him a "boss." Baogen's complexity reminds one of the workers' dual self-image that Lisa Rofel discusses in her chapter in this volume, in which their entrepreneurial past and/or future plays as crucial a role in identity construction as their current factory employment. At the same time, in his unwillingness to be bought off with a substitute and his readiness to wait for the real thing to be completed, Baogen also reminds us of the critic who says, "This is not the Shanghai I was promised—let us make the reality fit the name."

It is no accident that real estate is the key to Baogen's complicated positionality, even as it comes as no surprise that the movie ends with Baogen and Jinfang looking skyward, standing on the construction site for the building that will eventually contain his apartment. Although the now-you-see-them, now-you-don't quality of migrant laborers as New Shanghai People is delineated in part by perceived skills and educational level and in part by employment type, location and mode of residence also work to stratify new arrivals to the city. A Shanghai *hukou* certifies the right to be considered a legitimate long-term resident of the city, and it is significantly easier for white-collar employees and high-net-worth individuals to acquire; one's literal and figurative "place" in the city is established by residence in a renovated mansion in the old French Concession, a cramped apartment in the city's "lower corner," or something even smaller and more distant from the center. The association of the New Shanghai Person with questions of residence is confirmed in the frequent appearance of the term on websites concerned with real estate and residential policy: From their profiles on the Internet, one could be excused for thinking that the single issue preoccupying New Shanghai People is not their duty to participate in a world-historical reimagination of the city, but rather the task of locating a reasonably priced apartment.

Even as each New Shanghai Person is defined in part by the relations that he or she is able to enter into with the Shanghai real estate market, that market itself, once conceived as a process operating on the logic of capital, is crucially dependent on New Shanghai People as a cohort more generally. For an individual landlord renting a single apartment—and we note that Jinfang charges Baogen rent even though he is a relative!—an influx of newcomers to the city is welcome as a means of maintaining or increasing the rent that one can charge. Since the birthrate among city residents is below replacement value, in the long run in-migrants and immigrants are necessary to keep demand for housing high. In fact, the city's population grew by four million between 1992 and 2002 primarily on the strength of this in-migration and immigration (Dai 2003a: 210). Once we generalize beyond the individual landlord, and begin to consider corporations that invest in real estate, the logic becomes even starker. Investments of capital in various aspects of the real estate market cannot be justified by high rents alone; they depend on continuing increases in those same rents into the foreseeable future. Should the flow of in-migrants and immigrants financially positioned to rent or purchase apartments decrease, the construction boom will quickly result in a relative oversupply from an investor's perspective.[5] For all the transcendent significance of the new arrivals to the city as agents for broadening the city's vision and reconnecting it to its specifically national historical mission, is it possible that their simple need for a place to live is actually just as crucial to the city's continued development and prosperity?

Entrepreneurship and New Subjectivities

Historically speaking, the real estate market has in fact been one of the primary drivers of the city's development (Des Forges 2007: 19; Meng 2006). *New Shanghai People*, seemingly so preoccupied with an idealist reimagination of the Shanghai Person, turns out to have a close connection with these materialist dynamics: It results from a conference hosted by the Shanghai Zhengda Research Center, and it includes an introduction and conclusion by Dai Zhikang, the founder and CEO of the Shanghai Zhengda Investment Group, with strong interests in real estate and the stock market. This connection is nowhere more striking than in the essay by Chen Sihe, "Shanghai People, Shanghai Culture, and Shanghai Intellectuals." Like many Shanghai intellectuals, Chen is impatient with the nostalgia for the 1930s that has become so powerful by the end of the twentieth century. Where Hanchao Lu (2002) has noted the specific functionality of this nostalgia as a means for Shanghai residents to optimistically orient themselves in a new global economic order, keeping the underside of 1990s development out of mind in the same way that their nostalgia masks the uglinesses of the 1930s, Chen sees the focus on vulgar material culture characteristic of this nostalgia as a fundamental limitation (2003: 84–90). What is necessary for Shanghai's future development is not a modern culture that flows unproblematically from modernization at the material level, and thus itself may be viewed in material terms (expensive department stores, flashy nightclubs), but rather a transcendent culture of critical engagement that is able to call into question the falseness and banality of mass modernity.

It will surprise no one that a prime constituency involved in creating, articulating, and sustaining this transcendental culture critique is intellectuals, particularly Shanghai intellectuals, whom Chen Sihe finds to be more ready to criticize the city in which they happen to be located than intellectuals anywhere else in China. But the intellectuals' allies in this project, hinted at early in the essay, and finally making a full appearance in the closing pages, may come as somewhat of a shock: "private entrepreneurs" (*minying qiye jia*), many of them arriving in the city from elsewhere. "Unlike previous generations of immigrants to Shanghai, and unlike the current *mingong* class, [new in-migrants since the 1980s] rarely maintain fixed memories of their hometown cultures, and typically promote a new non-regional culture. In this way, the new inmigrants have created a distinctive cultural group in Shanghai" (2003: 84).

What private entrepreneurs have instead of narrow attachment to hometown culture is a more broadly based "popular cultural memory" (*minjian jiyi*), tempered by "orthodox education" (*zhengtong jiaoyu*) that seems—although Chen does not state this specifically—to take the nation as a whole as its object of focus. As a result, entrepreneurs have the ability to bring a critical and constructive perspective to the problems of Shanghai's mass culture; Chen notes particularly their potential to play a supporting role in Shanghai cultural production that allows for more spontaneity than that provided by the state (2003: 94).

Although more closely focused and readier to cite specific examples, Bao Zonghao's essay, "Private Technology Entrepreneurs: New Shanghai People of the Information Economy Era," continues along the same trajectory, beginning with the following question: Given that everyone likes to think of him- or herself as a New Shanghai Person, which group should be considered most representative and have its influence expanded in order to spur the city's development? Bao then goes on to conclude that New Shanghai People should refer to those individuals who contribute to the construction and development of the city (even foreign nationals) and participate fully in all aspects of "a completely new city in the information age" (2003: 135–136). Any thought that the former condition might involve the workers who literally build the new city is quickly dispelled by the latter condition; the "important representatives" (*zhongyao daibiao*) of the New Shanghai People turn out to be high-tech entrepreneurs, and it is the "quality" of these entrepreneurs that is the key factor in the success of their startup enterprises (138, 140).

It is tempting to see this celebration of private entrepreneurs and flowery language about the historical mission of in-migrants to the city as relatively transparent philosophical justifications for the status quo. With the rise of a class of young urban residents (native and in-migrant alike) who are "enslaved by their mortgages" (*fangnu*), is it too convenient that an essay collection like this insists that their sacrifice is not in vain, as their presence in the city has a higher purpose?[6] Yet to understand *New Shanghai People* as merely a reflection or expression of global, national, and local market forces at work—to dismiss the insistent rhetoric of consciousness raising and the cautions that Shanghai people themselves still have work to do in order to accomplish the task that History has set them, not to mention Bao's use of the highly charged term *daibiao*, as nothing more than false consciousness—would be to miss a critical dimension of this volume, and the discourse on New Shanghai People more generally.[7] What other function might these appeals for a leap in consciousness on the part of Shanghai people have? Are they nothing but ghostly echoes of "Maoist voluntarism" and other past calls to revolution?

New Era, New Protocols?

Unlike Chen Sihe and Bao Zonghao, Zhang Rulun maintains his focus steadily on a broader assortment of Shanghai residents and their views, but his conclusion, that "the world-consciousness of Shanghai people must be raised" (2003: 64), subscribes to the same basic understanding of the city's potential for future development. This "world-consciousness" (*shijie yishi*) is an organizing structure for the particular views that Shanghai residents hold about the city, other regions, the nation, and the world, yet for the ordinary resident this structure exists only at the *subconscious* level (59). We can deduce the presence of this subconscious structure through certain consistencies in residents' consciously expressed views, but how would these residents themselves be able to raise a "consciousness" of which they are not yet conscious? Perhaps it is the function of exhortations like those found in this volume to overcome this contradiction, but a crucial ambiguity is written right into Zhang's recommendation:

Consciousness must be raised but the agent of this action is left unnamed. In Chen Sihe's explanation of why the government cannot be the sole agent working toward a new "healthy" Shanghai culture, we notice a similarly careful negotiation in that "spontaneous" (*zifa*) cultural forces among the people nonetheless must wait to be "unleashed" (*fahui*). The obscured subject of Chen's verb *rang* (to "allow" or "cause"), in his assertion that "we must consider how best to . . . have (*rang*) the masses consciously develop national and regional cultural traditions," seems to bestow agency on these masses with one hand even as it is taken away with another (2003: 94).

In the face of a development that is historically necessary yet must also be voluntarily chosen by those who will accomplish it, is there any role for critics and public intellectuals other than exhortation? If not, Hegel's continuing influence on these authors may indeed be stronger than that of Marx; for where Marx was more than willing to countenance specific political actions undertaken to work in concert with the movement of history, no such grassroots organizing—indeed, no concrete activity whatsoever—is recommended in *New Shanghai People*. The central drama here is not the workings of a materialist dialectic, but rather the process through which a mind—or a city—becomes conscious of itself as consciousness.

Exhortation requires an addressee: Through the heyday of the movement period, from "Resist America, Aid Korea" to the "Anti-Spiritual Pollution Campaign," that addressee was the people as a whole. Today, the state in its national and local incarnations continues to address the people generally with injunctions against specific actions such as littering, spitting, and the use of vulgar language in public, as in the mid-1990s "Seven Don'ts" (*qibu*) campaign organized by the Shanghai City Civilization Committee (it is likely that the schoolchildren who chase after Jinfang for jaywalking in *Beautiful New World* are participants in this campaign). Reading with this tradition in mind, one begins by assuming that the exhortations to revolutionize one's thinking and become New Shanghai People are aimed at the residents of Shanghai generally, but doubts start to creep in, especially as we remember Zhang's claim that the "world-consciousness" of most of the city's residents is not in fact *conscious*. Calls to "be more global" may reach fewer people than "don't spit." Is it possible that the audience imagined for these exhortations is more limited?

Here we will want to think about the uses of terms such as *min*, *minzhong*, and *minjian*. Bao Zonghao's essay provides a useful point of entry in his movement between the use of *siying* and *minying* to refer to enterprises that are privately run. *Siying* indicates "private" as distinguished from "public," and the word appears when Bao describes prejudice against private enterprise that must be transcended; *minying*, on the other hand, is "private" in that it is not a part of the state and can also be translated as "nonstate." These nonstate enterprises are in turn run by *minjian qiye jia*, entrepreneurs who belong not to the state but to the *minjian* sphere, which Mayfair Yang defines as follows: "a realm of people-to-people relationships which is non-governmental or separate from formal bureaucratic channels" (1994: 288) (in certain ways this is analogous to the civil society thought to have formed in Europe in the modern era).

Although the fabric woven of *guanxi* networks that Yang understands as *minjian* may initially seem quite distinct from the way in which *minjian qiye* (nonstate enterprises) is now used, not to mention distant from the *mingong* who construct the city in which the *minjian qiye* can operate, significant connections can be traced forward and back. Yang's example of the *xuetou* ("cave head," or organizer of unofficial performance tours) as a leader of a purpose-driven group organized along *guanxi* lines led her, in 1994, to speculate that market development in Reform-Era China could take a form specifically indebted to *guanxi* networks (303–304, 309–310). Bao Zonghao, on the other hand, lists as one of his few criticisms of early twenty-first-century high-tech startups the degree to which their organizational structure depends on concrete relations between individuals rather than on a rationalized general plan (2003: 143–145). It is not an exaggeration, then, to say that both the high-tech enterprises that Bao praises and the groups of migrant workers that most discourse on New Shanghai People overlooks took shape in the Reform Era out of this generative fabric of nonstate relations. Even before 1978, it was evident that *guanxi* networks were not egalitarian and uniform, but took their place in and in turn themselves constituted unequal power relations between individuals. Between the *minjian qiye jia* and the *mingong*, we see this inequality taken to an extreme as new social groupings begin to separate out and solidify, but I would suggest that the metaphor of a fabric is still useful, as long as we understand it not as an undifferentiated expanse, but as a weave that is gathered more or less tightly at different points as the forces of capital twist and draw it into shape. This fabric is not fully autonomous, of course; its gathers attach to the more "rigid" state structure at key points. Bao notes the extent to which technology startups are run by individuals whose "quality" is validated by their educational credentials or by the fact of their previous employment in universities and government agencies, and highlights the assistance provided to nonstate startups by state enterprises (137–138).

Twists in the fabric of the *minjian* do not necessarily stem exclusively from shifts in the distribution of capital, but could take the form of reflexive turns—changes in consciousness—as well. In formulating the concept of the *minjian* as a social sphere, Yang refers to the work of Elemér Hankiss on the nonstate sphere in 1970s and 1980s Hungary. Hankiss writes of a "second society" that differs from Western European civil society in that it exists merely "in itself" rather than "for itself" as a separate and independent social formation (1988: 34–37). The crucial question here is of self-consciousness and the potential for conscious action as a social group, rather than as a collection of mere individuals. To the extent that appeals to Shanghai residents to become conscious of themselves as New Shanghai People—with an associated set of potentials and duties that constitute a historical mission—succeed, and to the extent that these appeals treat the state as a set of material conditions rather than as a participant with its own full agency to bring to bear in this process, we can understand the success of these appeals as movements toward the development of a "for itself" consciousness out of the *minjian*, a social identity that is rooted in the "spontaneous" energies of nonstate actors.

According to Wang Defeng, the category of New Shanghai People does not yet qualify as a "cultural concept" (*wenhua gainian*); New Shanghai People are at present still merely a group of in-migrants. The goal of *New Shanghai People* is to assess whether this group will acquire conceptual significance at some point in the future (2003: 13). Yet, given the emphasis throughout on self-conscious realization of one's place in history, and the authors' persistent appeals (*huhuan*) addressed to this very "group of inmigrants," one begins to suspect that the aim of *New Shanghai People* is rather to *constitute* New Shanghai People as a self-conscious social group through repeated exhortation. Just as the city must live up to the name that it already has, those in-migrants and new generations of Shanghai natives will make themselves into New Shanghai People precisely through their affirmative response to this call. In that sense, the title of the volume functions as apostrophe, bringing the object addressed to life.

Finally, we can return to the question of the organization under whose auspices this text was produced, the Shanghai Zhengda Research Center, identified as a "nonstate research organization" (*minjian yanjiu jigou*). Rhetorical invocations of Shanghai's unique historical conditions and mission, not to mention the insistence on making the city's reality adequate to its name, can be understood as strategic moves to fortify Shanghai's position against a variety of potential competitors for talent and investment, whether the cities of the Pearl River Delta, regional centers such as Chongqing, or even the capital itself. The publications subsequently organized and supported by the Zhengda Research Center continue this advocacy of the Shanghai region as uniquely qualified to lead the nation's economic development.[8]

Yet the work of this research center has deeper implications, which become clear if its sponsorship is contrasted with the many shallower alliances concluded between cultural producers and entrepreneurs beginning in the 1990s, a phenomenon that Shuyu Kong has traced in careful detail (2005: 144–169). *Beijing Literature*, a periodical that she cites as a typical example, made up for a decline in state support and readership by reaching out to corporations for financial support, apparently offering a quid pro quo that went beyond the sale of advertising pages to include placement of the enterprises, their products, and even their upper management within the "content" portion of the journal:

> In the January 1999 issue of *Beijing Literature*, the back cover offers a full-page ad for Beijing Badaling Tourism Corporation . . . with a photo of General Manager Qiao Yu prominently positioned. On the inside back cover is another picture of Qiao along with one of his poems in classical style, and [a] short curriculum vitae stating that he belongs to the Beijing Writers Association and once published a collection of poetry. Not surprisingly, Qiao Yu's name also appears at the top of the list of [*Beijing Literature*] board members as Chairman of the Board of Corporate Directors for 1999. (158)

The parallels between Qiao Yu's prominence in this issue of *Beijing Literature* and Dai Zhikang's power as CEO of the Shanghai Zhengda Group to write both the introduction and conclusion to a volume of essays by well-known Shanghai academics are evident. But the differences are also significant: Although the confluence of entrepreneur and elite cultural producers in *New Shanghai People* reminds us of the innovative approaches adopted by literary journals in the 1990s to assure their financial security, establishing a *minjian* research center allows the entrepreneur and his or her corporation a chance not merely to passively receive praise, but to actively help set an intellectual agenda.[9]

In the case of the "coeditorships" that Kong discusses, she notes that "there is little evidence of corporate synergy between most literary journals and their sponsors"; the aim of the "coeditor" is simply to exchange economic capital for cultural capital (159). Here, by contrast, the Shanghai Zhengda Research Center acts as a representative of a group that as a result of this representation can begin to constitute itself in "for itself" rather than "in itself" terms, with a potentially different vision from that articulated by the state. Given the difficulties NGOs often face in gaining state approval and maintaining a degree of autonomy, one wonders if a *minjian* research group that depends explicitly on a corporation whose assumed goal is profit above all else might not paradoxically be freer to articulate a distinctive vision than a stand-alone organization explicitly devoted to social change. When Dai Zhikang ends his concluding remarks with the words "Here I am just putting forward a few thoughts beginning from what is required to realize development in my own area, real estate: building Greater Shanghai" (2003a: 210), is the thinness with which the profit motive is disguised precisely the condition under which far-reaching philosophical claims about the relationship between local identity, history, and the nation may most safely and effectively be made? If Georg Wilhelm Friedrich Hegel were living in Shanghai today, would he claim maximizing the value of his real estate holdings to be his primary motive for writing philosophical works, and enjoy greater leeway to call for new forms of subjectivity as a result?

Notes

1 On the "humanist spirit" debates, see McGrath (2008: 25–58).
2 On the "livability" question, see Zhu Hong (2001) and Zhu Shengjian (2001).
3 The two that owe the least to this paradigm are Xiong (2003) and Deng (2003): The former sees the rise of a new Shanghai identity as following in relatively straightforward fashion from the resumed in-migration to the city; the latter defines New Shanghai People in terms of their use of the Internet and, despite a late note of caution on the scams and garbage that can be encountered there, indicates no doubt on the question of whether Shanghai people are up to the task of making positive use of it.
4 On the trope of the *chenggong renshi*, the "successful individual" who converts temporary control over or access to state-owned resources into personal wealth as state enterprises go private, see Huang (2004: 109–110) and Wang Xiaoming and Li Tuo (1999).
5 The uncertainty and instability characteristic of this market are referenced several times in *Beautiful New World*: Jinfang and her friend Ah Hui have lost a great deal of their assets through failed investments in real estate, and as the developer points out to Baogen, a lot can happen in eighteen months—right now it is estimated that his prospective apartment will be worth ¥700,000 when completed, but who is to say that accepting ¥100,000 in cash today in return for giving up his right to the apartment is not a better decision? One expects that audiences are supposed to laugh at the smooth Shanghai developer attempting to put such an obvious scheme over on the naïve Baogen, but he is not entirely wrong about the potential for market volatility.

6 For a detailed analysis of the figure of the *fangnu* in contemporary cultural production, with a focus on Shanghai, see Liang (2010).
7 Jiang Zemin introduced the theory of the "Three Represents" in a series of speeches in 2000; in 2002 it was established as a principle at the 16th Party Congress.
8 Including, among others, *Rushi hou Shanghai de zhanlüe diwei* and *Changjiang bian de Zhongguo—Da Shanghai guoji dushi quan jianshe yu guojia fazhan zhanlüe* (Chen Xueming 2009).
9 Not all research organizations with close corporate connections are this intellectually ambitious: The Ruan Yisan Foundation for the Protection of City Heritage, which is tied to Shanghai Ruan Yisan Urban Planning and Design Company, Limited, for example, keeps a significantly lower profile (Shao 2013: 263–265).

PART II

STRUCTURAL

RECONFIGURATIONS

4

DAMS, DISPLACEMENT, AND THE MORAL ECONOMY IN SOUTHWEST CHINA

Bryan Tilt

The World Commission on Dams (WCD), an organization under the guidance of the World Bank and the World Conservation Union, published a landmark study in 2000, which concluded that, despite their significant contributions to human development over the years, dams have had deleterious impacts on people and ecosystems. More troubling, the WCD noted, was the conspicuous lack of meaningful scrutiny in scientific and policy circles regarding the effects of dams around the world (WCD 2000). The WCD report estimated that about fifty thousand large dams, which the International Commission on Large Dams defines as those greater than fifteen meters in height or having a storage capacity greater than three million cubic meters, exist worldwide (Scudder 2005: 2–3). As the world's most populous nation and its second-largest economy, China's role in this trend is startling: Currently home to half of the world's large dams, it adds dozens of new ones to its portfolio every year in order to generate hydropower, provide sources of irrigation water, enhance the navigability of river systems, and protect downstream communities against flooding.

As hydropower meets a larger share of China's escalating energy demands, it may also help to reduce the nation's reliance on fossil fuels; government agencies and private entities alike are pursuing alternative energy-development plans involving not just hydropower, but also wind, solar, wave, and biogas. The development of a so-called low-carbon economy (*ditan jingji*) is welcome news in a country where hundreds of thousands of people die each year from ailments linked to air pollution from fossil-fuel combustion (Economy 2004), where pollution-related economic losses cut into the nation's gross domestic product, and where aggregate levels of greenhouse-gas emissions are the highest in the world.

China's push for hydropower development has attracted national and international scrutiny, however, because of a range of ecological and social ills that seem endemic to this form of energy production: fragmentation of riparian ecosystems, species extirpation, water-quality degradation, and the displacement of individuals and communities unfortunate enough to live in the path of large-scale infrastructure development. When dams are built and reservoirs fill behind them, they displace the human beings who live there, flooding farmland, inundating homes, and changing lives forever. The effects can last for generations, as people cope with the consequences for their families' income, their way of life, and their sense of place and community (Scudder 2005). China's growing hydropower sector thus represents one of the key points of friction between the competing rationalities of economic development, energy production, and social welfare.

The southwest region, with its glacier-fed rivers and high-relief topography on the edge of the Qinghai-Tibet Plateau, is home to the major share of China's vast hydropower potential. At present, a thirteen-dam hydropower-development plan is under way on the region's Nu River, in the northwestern corner of Yunnan Province, an area renowned for its cultural and biological diversity. These proposed dams will have a total hydropower potential of twenty-one thousand megawatts, which is slightly more than the mammoth Three Gorges Dam. Should all thirteen dams in the cascade be built, the best estimates suggest that more than fifty thousand people will be displaced. This is part of a pattern of hydropower development and human displacement on an unprecedented scale. In 2004, Xinhua News Agency, the official media mouthpiece of the Chinese Communist Party, released a short report on research conducted by the Ministry of Water Resources, which concluded that the nation's dam-induced displaced population totaled at least fifteen million people, ranking first in the world by a wide margin (Yao 2004).

The Nu River originates on the Qinghai-Tibet Plateau at an altitude of more than 5,500 meters, before flowing southward through the province of Yunnan and continuing through Myanmar (Burma), where it forms part of the Thai-Burmese border and is known as the Salween (see figure 4.1). With a total length of 2,018 kilometers, the Nu winds through the Gaoligong mountain range, a series of deep gorges and glaciated peaks created by the collision of the Indian tectonic plate with the Eurasian plate approximately fifty million years ago. The region encompasses a variety of ecosystem types from north to south, including glacial scree, alpine meadow, alpine conifer, deciduous forest, pine forest, mixed forest, savannah, and riparian habitats (Xu and Wilkes 2004).

The Chinese call the river Nu by using a character that literally means "angry" but is actually a phonetic rendering of Nong, the name given to the river by the local Lisu ethnic group, one of China's many minority nationalities (Mertha 2008). But the name also serves as an apt metaphor for the controversy surrounding the proposed dam-development project in this region that is home to six thousand plant species, eighty species of rare or endangered animals, and twenty-two of China's fifty-five officially recognized minority nationalities, many of whom have long endured economic and cultural marginalization on the edges of the Chinese nation-state.

In this chapter, I examine hydropower-development issues on the Nu River with an eye toward illustrating how different constituent groups—including government agencies and hydropower corporations, as well as communities facing permanent displacement—view the issue of dam development and attempt to shape policy outcomes. Drawing on the notion of the moral economy, I argue that we can look at the Nu River controversy through the lens of competing moral visions about what outcomes are proper, just, or otherwise desirable. On one hand, government policy and rhetoric emphasize the imperative to develop alternative energy sources to decrease reliance on fossil fuels, amid the backdrop of reforms in which unprecedented economic growth requires a seemingly endless supply of energy, even as the nation is reorienting itself from socialist ideals to a market economy. On the other hand, hydropower development causes the displacement and resettlement of thousands of people belonging to historically marginalized groups who stand to lose access to the agricultural land and forest resources that have supported their livelihoods. This raises key questions about rural property-rights regimes and the new ways in which local farming households and communities are brought into the sphere of the market economy as the nation veers away from Marxist and Maoist ideals. My analysis draws on a review of policy documents from central and provincial government agencies and hydropower corporations, along with surveys and interviews conducted in 2009 with the members of more than four hundred households in the Nu River Valley.

The Moral Economy of Water Resources in the Nu River Valley

Morality, and the normative judgments that stem from it, underpins all aspects of society, from legal institutions and political structures, to economic relationships and livelihoods. Precisely what is moral relates to both individual and societal values. David Graeber (2001: 1) has suggested that there are several broad streams of social theory about the notion of values, including "values" in the sociological sense, denoting shared concepts of what is ultimately good, proper, or desirable in human life, and "value" in the singular, economic sense, which relates to the degree to which objects are desired, usually measured by how much others are willing to give up in order to get them.[1] Both of these ways of constituting value are important threads in the story of Chinese hydropower development.

In his article "The Moral Economy of the English Crowd in the 18th Century," the historian E. P. Thompson defined a moral economy as a "popular consensus . . . grounded upon a consistent traditional view of social norms and obligations, of the proper economic functions of several parties within the community" (1971: 79). The central premise is that economic and political interactions are shaped by norms that are rooted in much deeper cultural and historical precedents. Scholars have used this model to examine people's ideas about rights to subsistence and agricultural production (Scott 1976); expectations of the state in the provision of social services (Sivaramakrishnan 2005); the role of market economies in producing and distributing commodities (Griffith 2009); and competing goals and priorities in water management (Wutich 2011).

As a vital element in sustaining life, and as the cornerstone of many economic and cultural activities, water is a commodity that lends itself well to this kind of analysis. A society's political and economic choices about water use and allocation reveal a great deal about its core values and about how those values are constructed and maintained through everyday forms of practice. Indeed, when new economic, political, or environmental events violate people's sense of what is just or proper, civil unrest or mass social movements may result (Edelman 2005). This underscores the fact that a shared sense of propriety, which can be fragile to maintain, is at the core of human social relations.

In China, the idea of a moral economy carries particularly effective explanatory power, because the country is currently undergoing far-reaching social and economic reforms that are pushing the boundaries of traditional forms of moral consensus. As the Chinese Communist Party has liberalized the nation's economy over the past three decades of Reform and Opening Up, it has also gradually reduced the scope of its administrative power, thereby increasing the space within which civil society organizations may operate (O'Brien and Li 2007; Tilt 2010; Weller 2005). This is especially important in the environmental arena, where a new Environmental Impact Assessment Law, promulgated in 2004, mandates public hearings for major development projects. Environmental lawsuits are increasingly common, as are environmental nongovernmental organizations (NGOs), which number in the hundreds (Yang 2005).

Under the constitution, the PRC considers itself a "unified, multiethnic state" (*tongyi duo minzu guojia*). In addition to the dominant Han majority, which constitutes about 91.5 percent of the nation's population, there are fifty-five "minority nationalities" (*shaoshu minzu*) that received formal recognition by the central government following an ethnic-identification project conducted between 1950 and 1956. The aim of the project was to classify each minority group according to Marx's schema as either primitive, slave, feudal, bourgeois-capitalist, socialist, or communist (Harrell 1995). The government borrowed its classificatory criteria from the Soviet Union's system, developed by Stalin, which included a common territory, language, mode of subsistence, and "psychological makeup." In practice, the process of affording recognized minority status to some groups while denying it to others was politically charged, since it was tied to the establishment of ethnic regional autonomy for governance in minority areas and to the provision of social and economic benefits (Wang and Young 2006).

These minority nationalities, which numbered 114 million as of the 2010 census (approximately 8.5 percent of the total national population), represent a particularly vexing development problem for the government. On one hand, the fact that they are viewed as "backward" (*luohou*) provides normative justification for targeted development, economic assistance, educational subsidies, and national welfare policies. On the other hand, high concentrations of minority nationalities are perceived by many citizens as a barrier to actually achieving development at a level consistent with the national average.

Various places with high concentrations of ethnic minorities carry the designation of "autonomous region" (*zizhi qu*), "autonomous prefecture" (*zizhi zhou*), or "autonomous county" (*zizhi xian*), but the ability of officials within these entities to practice self-governance or to influence central policy remains extremely limited (see Rossabi 2004). Tibet, for example, is recognized as a provincial-level autonomous region (*zizhi qu*), but the potential for a secessionist movement causes the central government to keep a tight rein on the region. Moreover, China has been reluctant to accord its minority nationalities "indigenous" status under such frameworks as the UN Convention on Biodiversity, which recognizes the rights of indigenous people for autonomy and self-determination.

Northwest Yunnan's Nujiang Prefecture, where eight of the thirteen dams are slated to be built, was established in 1954 as an "autonomous prefecture" for the Lisu people. The Lisu language is Tibeto-Burman in origin and has about six hundred thousand speakers in Southwest China and thousands more in Burma and Thailand (Ethnologue 2009). However, ethnic boundaries in the Nu River Valley are often not so clear-cut. Individuals from different ethnic backgrounds—including Tibetan, Nu, Lisu, and others—have intermarried for a long time; during ethnographic fieldwork, when I inquired about the ethnic identity of a given individual, I routinely heard a long pause, followed by an intricate recitation of genealogy: "My father is Lisu, but my mother is Tibetan, and many members of my extended family are Nu." As a result of this prolonged, intimate contact between ethnic groups, facility in two or three languages is common among local residents.

Like many of the highland ethnic groups in Southwest China, the Lisu historically practiced swidden agriculture. But the rise in population over the past several hundred years, coupled with the market-based agricultural policies of the Reform period, induced them to practice sedentary agriculture. They grow a variety of grain crops, including rice, corn, barley, and buckwheat, and a wide array of vegetable crops. Agricultural fields are carved out of incredibly steep hillsides or situated on alluvial fans where tributaries to the Nu River have deposited rich, fertile sediment. Maintaining this cultivation pattern requires a tremendous amount of human labor, which the anthropologist Marco Clark has referred to as a "delicate balance between gravity and human determination" (2009: 23). The push for hydropower development has placed the Lisu and other Yunnan minority groups at the center of a growing controversy over national priorities and policies regarding energy production.

State Agencies and Hydropower Corporations

As with many sectors of the Chinese economy during the Reform Era, the political economy of hydropower development entails a blending of state control and private investment. The rights to develop water-management infrastructure have been turned over to private interests (albeit those headed by economic elites with close government ties), while the overall goals and priorities of water management remain under the control of the state. The National Development and Reform Commission, the Ministry of Water Resources, and other government agencies have the ability to approve particular projects and block others, or to provide tax incentives and favorable financing to projects that further the economic goals of the state.

In 2002, the State Power Company of China was dissolved and its assets distributed among eleven limited-liability stock corporations, five of which became responsible for power generation.[2] One of these "big five electricity giants" (*wu da fadian jutou*), China Huadian Corporation, holds a state-granted monopoly on the right to develop dam projects on the Nu River. Huadian enjoys close ties to the provincial government in Yunnan and exerts considerable influence over water resource development plans (Magee and McDonald 2009). Information about each of the thirteen dams in the Nu River Project is given in table 4.1, from Songta Dam in the north, which is located in the Tibetan Autonomous Region, to Guangpo Dam in the south.[3] Design and operation specifications differ considerably from one dam to the next. Some, such as the Songta and Maji dams, are very large (more than three hundred meters high) and will include reservoirs with enormous storage capacities, displacing thousands of people; others, such as Bingzhongluo and Liuku, will be "run-of-the-river" dams with minimal storage capacity and will thus displace far fewer people. Current estimates suggest that 51,079 people would be displaced, should all thirteen dams be built (Tullos et al. 2013).

Since 1978, Chinese leaders have followed a path of economic liberalization and privatization known as Reform and Opening Up (*gaige kaifang*). Throughout the Reform period, the nation's annual rate of GDP growth has been at or near double-digit figures, and this push for economic and industrial development has coincided with a steep increase in energy use, powered largely by fossil fuels. On the eve of Reform and Opening Up in 1978, China was a poor, inward-looking, agrarian country. Driven by an exponential increase in manufacturing between 1980 and 2010, the nation's total annual energy consumption increased by a factor of five, to reach the equivalent of 3.25 billion tons of coal being burned each year (Liu 2012). Even conservative projections suggest that China's energy consumption will continue to rise in the coming years, if at a slightly slower rate, as its economy matures.

Coal, by far the most abundant energy source in China, provides more than two-thirds of the nation's energy supply, with industries, power plants,

Table 4.1 Design Specifications for the Thirteen Proposed Dams in the Nu River Project

Dam Name (Listed from North to South)	Height (m)	Installed Electrical Capacity (megawatts)	Reservoir Storage Capacity (million m³)	Estimated Population Displaced
Songta	307	4,200	6,312	3,633
Bingzhongluo	55	1,600	13.7	0
Maji	300	4,200	4,696	19,830
Lumadeng	165	2,000	664	6,092
Fugong	60	400	18.4	682
Bijiang	118	1,500	280	5,186
Yabiluo	133	1,800	344	3,982
Lushui	175	2,400	1,288	6,190
Liuku	36	180	8	411
Shitouzai	59	440	700	687
Saige	79	1,000	270	1,882
Yansangshu	84	1,000	391	2,470
Guangpo	58	600	124	34
Total		21,320	15,109	51,079

Sources: He, Feng, and Hu 2007: 147–148; Magee and McDonald 2009.

and households consuming 1.5 billion metric tons per year (Smil 2004). The emissions from such activities contribute to widespread environmental degradation and public health risks at home, and to global climate change. Although its dependence on coal and other fossil fuels will persist for many years, China is investing heavily in alternatives, which policy makers refer to as "clean energy" (*qingjie nengyuan*), "green energy" (*lüse nengyuan*), or "low-carbon energy" (*ditan nengyuan*). By 2020 China has committed to both reducing overall carbon emissions to 40–45 percent below 2005 levels, and reducing carbon intensity, or the carbon emissions required to produce one unit of GDP, from a current level of 2.7 tons per $10,000 of GDP to a target of 1.2 tons. Hydropower, for all of its social and environmental ills, is seen by many officials in China as a potentially "clean" source of renewable energy and now accounts for more than 16 percent of total electricity production (Liu 2012).

At the same time, large-scale dam construction is also viewed as a development path that can make considerable contributions to national and regional economic growth. Chinese scientists and policy makers have explicitly drawn inspiration from the U.S. example, noting in particular how the growth of California and other Western states was made possible by the huge expansion in dam construction in the 1930s, 1940s, and 1950s (Wang, Yu, and Li 2008). These dams, many of which began under the Works Progress Administration of the Great Depression as job-creating strategies, made both new water sources available for expanding municipalities such as Los Angeles and new sources of electricity available to meet growing industrial and consumer demands.

Hydropower development is also a key part of China's "Develop the West" strategy (*Xibu dakaifa*), enshrined in the nation's Tenth and Eleventh Five-Year plans for economic development (2001–2010), the goal of which is to narrow the economic and social disparities between the prosperous east coast and the relatively impoverished western regions, including Yunnan. Moreover, the Twelfth Five-Year Plan (2011–2015) explicitly cites the southwest region as a key "hydropower base" (*shuidian jidi*) within the nation's overall energy-development strategy.

The hydropower resources of this region are vast; most of the nation's great rivers—the Yellow, the Yangtze, and the Mekong, among others—have their headwaters here in the expansive, high-altitude, and arid interior. It is precisely these regions, moreover, that have been largely left behind in China's rush to develop its east coast, leading political leaders and hydropower officials to argue that dams can contribute both to the nation's energy supply and to jobs, revenue, and improved infrastructure in Yunnan. Officials in county- and provincial-level government agencies hope that these projects can improve local transportation infrastructure, further develop communications technology, provide employment opportunities, spur new investment, and even improve educational levels (Li 2008).

But the challenges of bringing economic development to this region—one of the nation's poorest—are considerable. At the national scale, no province with a minority population greater than 10 percent is listed among the ranks of high-income or middle-income provinces (Wang and Hu 1999). Beyond simple economic measures, the United Nations Development Programme (UNDP), for example, uses a Human Development Index (HDI), which includes measures of economic productivity, life expectancy, and education, to identify development needs. In the most recent HDI calculation, Yunnan ranks twenty-ninth out of thirty-one provinces and administrative regions (UNDP 2008: 22). In Nujiang Prefecture, minority nationality people account for 92.4 percent of the 520,600 residents, and all four counties in the prefecture are designated national-level impoverished counties, which qualifies them to receive subsidies from the central government. This is a common trend in Reform-Era China; measured in a variety of ways, income and wealth inequality has continued to grow throughout the Reform Era, widening gaps between rural and urban communities and between individuals within communities.

Many of the official government documents on the Nu River Project describe the dams as a "poverty alleviation weapon" (*fupin wuqi*) or as one component of a broader "poverty alleviation strategy" (*fupin zhanlüe*) in a region that is perceived by mainstream Chinese society as culturally and economically backward. The government of Nujiang Prefecture, which is accustomed to relying on central government subsidies as a main revenue source, is strongly in favor of the projects, which would generate an estimated RMB 36 billion per year in revenue, one billion of which would go to Nujiang Prefecture, effectively increasing the prefecture's revenue flow by a factor of ten (Magee and McDonald 2009: 50).

Careful inspections of the planning documents, however, reveal that the vast majority of electrical power will be sent eastward to coastal cities, mostly in the Pearl River Delta region, where industrial, commercial, and residential demands are high. This plan, officially termed "Send Western Electricity East" (*xi dian dong song*), will be facilitated by high-voltage direct-current (HVDC) transmission lines: cutting-edge technology that Chinese engineers helped to develop. Thus, existing economic disparities between regions are likely to be exacerbated by the Nu River dams (Magee 2006). Furthermore, the economic feasibility of the projects is questionable. Construction will require a significant outlay of capital for material, labor, and support infrastructure such as roads, plus more than RMB 3 billion for resettlement costs alone. Government reports estimate the commercial value of hydropower produced by the projects to be RMB 16 billion per year, though the lifespan of the projects and future electricity prices are highly uncertain (He, Feng, and Hu 2007).

Local Communities, Displacement, and Resettlement

When dams are built and reservoirs fill behind them, the individuals and communities who live in these catchment areas must undergo displacement and resettlement. Their agricultural land is inundated, their homes are submerged, and their lives and livelihoods are changed forever. For villagers in the Nu River Valley, displacement compounds the long-term effects of economic and cultural marginalization. In recent years, the region's integration into the market economy has led to the expansion of agricultural fields for the production of commodity crops and a consequent loss of forest land as villagers struggle to make a cash income in order to educate their children and access other social services that were previously provided by the state (Xu and Wilkes 2004). Dam construction thus represents the latest in a series of major social and environmental changes in the Nu River Valley in recent years.

To understand the perspective of local community members on hydropower development, a group of U.S. and Chinese researchers conducted household surveys in Nujiang Lisu Autonomous Prefecture in 2009. Our group surveyed 405 households in two counties (Fugong and Lushui), encompassing 13 townships and 20 villages, including both upstream and downstream communities related to four proposed dam sites: Maji, Lumadeng, Yabiluo, and Lushui. Household members were asked to provide information on a range of issues related to income, livelihood activities, ethnic and cultural identity, community participation, and education. In addition, the research group conducted qualitative interviews with a random subsample of forty-eight households that participated in the surveys, asking questions about the perceived benefits and costs of dam construction and the means available to villagers for coping with potential changes to their lives and livelihoods.

Nujiang Prefecture is one of the poorest regions of China. Within the survey sample, the median household income was approximately RMB 3,700 (USD 587) per year, and 70 percent of households reported receiving poverty alleviation subsidies from the central government. The primary economic activity is farming for subsistence and market exchange, supplemented by the gathering of non-timber forest products such as mushrooms and herbs, a practice that has already been curtailed by the establishment of nature reserves that prohibit economic use. Dam development is occurring simultaneously with massive road-construction projects that will soon allow easy passage from the Nu River Gorge eastward to Deqin, Shangri-La, and Lijiang, which are already popular tourist destinations. In the midst of this construction boom, the subsistence economy based on crop production and wild resource harvesting is gradually giving way to wage-labor jobs, migration to cities and towns, and tourism-related work. The Nu River dam projects thus represent just one piece—albeit an important one—of a recent pattern of massive socioeconomic change.

Villagers' understanding of and attitudes about dam development in the Nu River basin are quite varied. More than half of the village households included in the sample lacked any systematic information about specific plans for hydropower development, and they knew little about the scale of the dam projects that would affect them and the time frame within which such projects would be built. Despite the mandate for public hearings, as stipulated in China's Environmental Impact Assessment Law, most villagers had not attended such hearings. For example, in Xiaoshaba Village, where residents had recently been resettled upstream of the Lushui Dam site, a majority of residents reported a general lack of information about the projects; their primary source of information was from other villagers. A major factor influencing knowledge about the dam projects, and participation in public hearings, appears to be facility in standard Mandarin Chinese. Most residents in the Nu River Gorge speak either a minority language or a local dialect of Chinese, making it difficult for them to access Chinese-language materials from government authorities or the media.

Despite this fact, many villagers expressed support for the construction of dams in general across the nation, and many also supported the Nu River projects specifically. Most villagers cited electricity generation and economic growth as the most salient benefits that would likely come from dam construction. One man, a farmer in Lushui County, said, "It [dam construction] can turn a natural resource into an energy source. Flowing water becomes economic income that strengthens the nation's economy and builds the nation's prestige." At the local level, many villagers anticipated being employed in construction and regarded this as a means of gaining wage-labor income without having to migrate out of the community to look for work.

Nevertheless, most villagers recognized that such benefits would come at a significant cost. Many expressed concern about the inundation of their farmland and the loss of their homes, which, at a median value of RMB 30,921, constituted the most valuable asset for nearly all families. The consequences for them and their families could be felt for generations, particularly if access to land—long considered the primary social safety net in rural China—was compromised. One forty-year-old farmer in Lushui County, reflecting on the possibility of resettlement in the near future, remarked, "Although villagers will receive compensation, they'll lose their land and that will mean their children will be landless peasants."

Villagers were asked how they would cope with the effects of the dam projects on their livelihoods, and many of their responses reflected a general sense of fatalism, that "nothing could be done" (*meiyou banfa*). As one fifty-two-year-old farmer said, "Whatever the government says is what we must do." On the other hand, villagers' responses also pointed to a sense of trust that government officials and policy makers would compensate them fairly for their lost land and homes. One forty-six-year-old woman commented that "the common people don't have any means of coping by themselves, but I believe the government will provide a plan for us." This finding is in line with current research in political science that shows that rural Chinese villagers often trust central government authorities far more than they trust local government cadres, who are commonly viewed as opportunistic, self-serving, or even corrupt (Li 2004). It remains to be seen whether such trust is well placed.

Despite these concerns, there is little evidence of a grass-roots oppositional movement in the Nu River Valley (Litzinger 2007). Such reticence is attributable to the challenges of obtaining information about how projects are proceeding, a weak local capacity to mount a campaign, and the high political risks involved in any overt opposition strategy in China. As I discovered, even answering interview and survey questions about the dam projects can be considered politically sensitive (see also Magee and McDonald 2009). There are visible opposition movements, to be sure, but most such campaigns remain the purview of national or international NGOs, some focused on the fate of the migrant population and others concerned with threats to the local ecosystem and its biodiversity (Tilt 2015).

Mao, Marketization, and Rural Property Rights

Any attempt to understand the effects of hydropower development on local people must be grounded in an analysis of the changing political economy in Reform-Era China, which David Harvey (2005: 120) has characterized as "a particular kind of market economy that increasingly incorporates neoliberal elements interdigitated with authoritarian centralized control." While I am skeptical about applying the term *neoliberal* to contemporary China, since it connotes a combination of liberal market policies, an eschewing of central economic planning, *and* at least a rhetorical emphasis on individual freedom and liberty (Harvey 2005), I think the concept can shed some light in this case. In particular, Aihwa Ong's (2006: 3) analysis suggests that neoliberal governance strategies in China and elsewhere tend to be "recast as nonpolitical and nonideological problems that need technical solutions."

The dismantling of the State Electric Power Corporation in 2002, and the distribution of its assets to various state-owned enterprises and shareholder corporations, constitutes an application of market logic to China's electric-power sector. In a fundamental sense, displacement and resettlement to make way for the dams in Yunnan represent villagers' head-on encounter with the market logic of Reform-Era China. Moreover, when people are in the way of such progress, they are dealt with through technocratic, economic, or legalistic methods that often come in the apolitical guise of simply the "workings out of the market" (Oliver-Smith 2010: 3). The emphasis on market-driven development in the electricity sector is a crucial tactic employed by the "big five electricity giants," since market solutions are pursued with a kind of utopian idealism in contemporary China. But it would be a mistake to ignore the roles played by various state agencies too. In fact, hydropower development in contemporary China represents a blending of the most troublesome elements of two systems: a capitalist market with a hunger for constant expansion and the means to externalize costs onto people who lack political power, coupled with an authoritarian state capable of removing many of the obstacles that stand in its way.

One of the most intractable problems with resettlement policy in China relates to rural land tenure. During the Maoist period, agricultural production was controlled by the state via a network of communes, production brigades, and production teams overseen by party cadres. This system became the basis for economic and social life in rural China for nearly three decades. The Household Responsibility System, introduced in the early 1980s, granted farming households the right to make their own production decisions and earn residual income from their land by selling crops on the market, which was subject to fewer and fewer government restrictions. This effectively reestablished China as the world's largest smallholder agricultural system based on millions of households pursuing small-scale, intensive farming.

But vestiges of the Maoist system remain intact, and the current property-rights regime that governs access to rural land continues to be one of the greatest headaches for political officials and rural households alike. The Household Responsibility System gives farming households access to cash income through agricultural sales. Decisions regarding crop selection, cultivation, and market distribution are all made at the household level, and economic risk within China's rapidly changing market economy is also assumed at the household level. But security of future tenure rights and transferability to a third party are both major issues of concern in rural areas. In contrast to urban land, which is owned by the central government, rural land rights are vested in rural collectives at the level of township, village, or production cooperative, an arrangement that is codified in the constitution of the PRC.

This amounts to an incredibly complex and ever-shifting property-rights regime that has been characterized by a "deliberate institutional ambiguity" (Ho 2001; Rozelle et al. 2005). Individual farming households are typically granted certificates that give them use rights, along with the right to sublet land within the term of their lease, but not full ownership rights over two types of land leased from the rural collective: "responsibility land" (*zeren tian*) and "contract land" (*chengbao tian*). The length of rural land leases has increased dramatically since the 1980s and now spans seventy years for "responsibility land," although the collective retains the right to appropriate land within its jurisdiction when necessary. On contract land leased from their collectives, farmers now enjoy "indefinite" lease periods; the hope of government planners is that this relative sense of long-term security will help to promote specialization, increased land transfers, and economies of scale in the agricultural sector. This shift in agricultural governance has greatly increased farming income, a welcome trend for most rural people, but it also creates considerable risk for individual farming families, who must meet their own economic needs as the state provides less security and fewer services than during the socialist period.

We might briefly consider the implications of this institutional ambiguity for dam-related resettlement projects like the ones under way in the Nu River Valley. On September 1, 2006, the State Council, China's chief legislative body, passed the "Regulations on Land Acquisition Compensation and Resettlement of Migrants for Construction of Large and Medium Scale Water Conservancy and Hydropower Projects" (State Council 2006). These regulations represent a dramatic step forward from China's historically inadequate compensation structure. For example, the regulations stipulate that resettled households are to receive sixteen times the value of their average annual income, in addition to compensation for housing of the "same scale, same standard, and same function" as what they previously had (Brown and Xu 2009). The regulations also stipulate that compensation for resettlement must be included in any project budget and that potentially resettled people have a right to know about development plans affecting them and to participate in the process through public hearings.

For local villagers, the long-term consequences of the Nu River Project will likely depend on how well these new compensation policies are implemented and enforced, which remains to be seen. Thus far, the trouble appears to lie in the implementation of regulations. Just upstream from the Liuku Dam site, where preparatory work is under way, 144 households from Xiaoshaba Village were relocated to New Xiaoshaba Village in 2007. Although public hearings were held in 2006, most villagers reported feeling intimidated and effectively shut out of the decision-making process. More than a year after resettlement, no steps had been taken to allocate new farmland to the resettlers, and villagers reported a lack of long-term support programs such as job training (Brown and Xu 2010).[4] Much of the land that will be inundated by the dam's reservoir is low-lying, fertile rice paddy on alluvial fans adjacent to the river; such land represents generations of agrarian labor and constitutes a major share of household income for local villagers.

More problematically, compensation policy is based on an extremely narrow definition of ownership. Far from homogeneous, property rights may encompass a wide spectrum of entitlements, from informal access rights on one end, the right to earn a living by farming in the middle, and to full ownership and transferability on the other end (Ribot and Peluso 2003). But the PRC, for all its strides toward a liberal market economy, still effectively recognizes the middle part of the spectrum only. When villagers are forced to move in order to make way for a development project, the law requires that they be compensated, but only for lost income from agricultural production, not for lost ownership of the land and whatever future value or productive capacity it might have embodied. The law does not recognize ownership or salability at one end of the land-rights spectrum, nor does it recognize, at the other end of the spectrum, informal resource-harvesting rights on forest land that has historically been communally managed.

To provide just one example of how the land-rights regime affects displaced people, villagers in the Nu River Valley have historically relied on a diverse assemblage of resources over which they hold no formal title at all. Many households still gather mushrooms, herbs, and firewood on forest land under informal agreements with their relatives and neighbors, using these products as part of their daily subsistence and sometimes selling them on the local market to supplement their income from farming. Villagers also grow corn and other row crops on deforested hillside land that does not belong to them at all; in surveys and interviews, it was not uncommon for villagers to acknowledge having planted more than ten *mu* (1 *mu* = 1/6 acre) of corn as a cash crop while officially reporting total household land holdings amounting to only half of that amount. Many forms of land use in rural China are, strictly speaking, extralegal, but they nevertheless constitute a significant part of household economic activity, particularly for vulnerable people whose income scarcely rises above mere subsistence.

To frame the problem in Marxist terms, we might be tempted to call it "dispossession" or "expropriation," both processes that are part of the "historical tendency of Capitalist accumulation" in Marx's view (1978: 436). After all, the story of capitalist development has long entailed pushing peasant households onto the conveyor belt of the global market system, using their land for "higher" economic purposes and putting their labor to work in cities and towns (Graeber 2001: 78–80). But in truth the Nu River case may represent something even more problematic: the requisitioning of assets that vulnerable people rely upon but, by virtue of a land-rights regime with divided loyalties between Marx and the market, hold no security of title over.

In *The Mystery of Capital*, the economist Hernando de Soto (2003) observes that the single greatest obstacle faced by most of the world's poor is a lack of formal property rights. One need not be a proponent of private property to see that clearer and more transparent property rights are the basis of a stable and just economic system. Well-defined property rights of whatever type—whether vested in the individual, the community, or the state—are the ultimate source of wealth, especially in rural areas where land constitutes people's most productive asset and their most abiding source of future security. One might reasonably ask whether powerful interests such as government agencies or hydropower corporations actively seek to undermine the livelihoods of the poor through a process of dispossession. The short answer is: Perhaps. But such overt tactics are not usually necessary, because the existing, ambiguous property-rights regime already works in favor of these powerful political and economic actors.[5]

Implications for the Moral Economy of Water and Energy

The status of the Nu River Project changes constantly, based on the prevailing political winds, and the project remains a focus of controversy and contention both domestically and abroad.[6] After considerable opposition from Chinese and international conservation organizations, along with NGOs concerned about the status of displaced people, Premier Wen Jiabao ordered a temporary halt to all Nu dams on February 18, 2004. This was considered the first serious test of China's new Environmental Impact Assessment Law, which was promulgated in 2003. Under Premier Wen's advisement, the National Development and Reform Commission, along with the State Environmental Protection

Administration, conducted a review of hydropower development on the Nu. Early in 2006, officials decided to allow construction to commence on a scaled-down version of the projects, beginning with four of the thirteen dams: Maji, Yabiluo, Liuku, and Saige. Two of the four, Liuku and Saige, were set to begin construction during the Eleventh Five-Year Plan period (2006–2010), according to a document released by the National Development and Reform Commission, entitled "The Plan on the Development of Renewable Energy during the Eleventh Five-Year Plan Period" (Brown and Xu 2010).

But plans were stalled yet again by Premier Wen in April 2009, who declared somewhat obliquely that authorities should "widely heed opinions, expound on [the plan] thoroughly and make prudent decisions" (Shi 2009). Speculation abounded that Wen, who has an academic background in geology, was concerned about potential seismic hazards in the aftermath of the Sichuan earthquake in 2008.[7] However, by early 2011, top officials had again decided to move forward with the projects; Shi Lishan, the deputy director of the Energy Department, was quoted in *The China Daily* as saying, "I think it's certain that the country will develop the Nujiang River" (*China Daily* 2011). On January 23, 2013, the State Council announced that Songta Dam, located in the upper reaches of the watershed in the Tibet Autonomous Region, would officially begin construction during the Twelfth Five-Year Plan period (2011–2015). A pre-feasibility study was approved in late 2014, moving the Songta Dam one step closer to beginning construction. Four more dams—Maji, Yabiluo, Liuku, and Saige—are also undergoing feasibility studies and will likely move forward as part of the scaled-down version of the Nu River Project (International Rivers 2013).

What are the implications of the Nu River hydropower projects for understanding the moral economy of water and energy development in contemporary China? As the Nu River case demonstrates, morality is not constructed neatly from the top down; rather, it is emergent and constantly negotiated. In fact, underlying the Nu River case are multiple, fragmented regimes of morality, borne of individual struggle, institutional actions such as law and policy, and public discourse at regional, national, and international scales. Key government agencies, including the National Development and Reform Commission and the Ministry of Water Resources, view hydropower development as a means to achieve energy security and to reduce the carbon intensity of the nation's economy amid the continually rising demand for electrical power. They work with the "big five electricity giants," each with its own tangled list of subsidiary companies, to develop Yunnan's hydropower potential and distribute electricity eastward on the national grid. Improving energy security and reducing the nation's reliance on fossil fuels are significant and laudable goals with the potential of generating enough electricity to offset, and perhaps even render obsolete, hundreds of coal-fired power plants that are at the heart of the nation's air-pollution woes.

But while the story of hydropower development may begin with national energy plans and policies, it certainly does not end there. Most villagers in the Nu River region readily acknowledged, and even agreed with, the national and provincial governments' arguments about the need to develop hydropower for the good of the economy. Yet the fact remains that the benefits of hydropower development (namely, increased access to electrical power for industrial production) would accrue to comparatively wealthy eastern cities while the costs—including displacement, potential landlessness, and inadequate implementation of compensation policies—will be borne by extremely vulnerable agrarian communities. I have suggested that this fact represents an unresolved point of friction between multiple, often conflicting, moral visions about the future of the Nu River.

The question is where to go from here as the new, scaled-down plans for hydropower development move forward. If, as I have suggested, notions of morality are dynamic and emergent rather than static, then moral problems can be reenvisioned and more just solutions can perhaps be fashioned. I would like to reflect briefly on how the findings and analysis presented in this chapter can contribute to the debate about balancing hydropower development with the needs and livelihoods of local people. First, it is clear that any large-scale hydropower-development project should be implemented in such a way as to avoid damaging, or ideally even to improve, the living conditions of affected people. Even publications sponsored by the Chinese government underscore the importance of "improving the living standards of local minority communities" (Lu 2008).

A participatory approach that brings local individuals and communities into the process is critical. While most residents of the Nu River Valley lack both information about the dam projects and the opportunity to participate in public hearings, evidence from studies elsewhere in the region suggests that development projects undertaken by both state agencies and foreign NGOs in China are adopting participatory approaches with greater frequency, driven in part by central policy that mandates public hearings as part of the guidelines for complying with the Environmental Impact Assessment Law (Tilt 2015: 148–149). In countries around the world, the World Bank and other multilateral agencies involved in hydropower development have established clear guidelines for assessing the effects of dams on communities. The landmark report published by the WCD in 2000 emphasizes the application of a "rights and risks" approach to evaluating dam projects, which entails, among other things, "Self-determination and the right to consultation in matters that affect people's lives, the right to democratic representation of people's views on such matters, the right to an adequate standard of living, freedom from arbitrary deprivation of property" (200).

Such a framework, however, has been slow to gain traction in China, where governmental policies tend to be top-down. The WCD is an advisory body that lacks any enforcement mechanism, particularly in cases such as the Nu River where dam projects are funded by domestic government agencies or corporations.

Second, if progress toward just compensation is to be made, it will require a consideration of what remains the most challenging issue in China's economic transformation: rural land tenure. Villagers in the Nu River Valley, like elsewhere in China, are granted usufruct rights over their land via the Household Responsibility System but hold no formal land title, which still belongs to the rural cooperatives, the vestiges of communal farming during the Maoist period. Many of the failings of current compensation policies have their roots in this form of institutional ambiguity: Villagers receive compensation based on lost agricultural income, while failing to capture the long-term value of agricultural land because ownership is ultimately vested in the cooperative, and also losing access to informal resource-gathering rights that comprise an important component of their livelihoods.

At present, it remains unclear exactly how many dams will be constructed on the Nu and according to what time schedule. The Nu River case has brought to the forefront a set of moral questions about how to balance energy production in a rapidly developing nation struggling to find alternatives to fossil fuels against the imperative to protect the livelihoods and rights of local people. Decision makers find themselves at a crossroads in which critical choices are being made about how to power China's economic future amid economic liberalization, escalating energy demands, and increasing socioeconomic inequality. While it is difficult to define what might constitute a "successful" outcome in a case as complex as this one, it is clear that success will entail a full and transparent accounting of the social costs of the Nu River dam projects and a commitment to hear and address the concerns of historically marginalized communities.

Notes

Funding for this research project was provided by the U.S. National Science Foundation's Human and Social Dynamics Research Program (Grant #0826752). I would like to recognize the contributions of other project personnel, including Philip H. Brown, Feng Yan, He Daming, Darrin Magee, Shen Suping, Desiree Tullos, and Aaron T. Wolf. I would also like to thank the following collaborators and students for their tireless work during data collection and analysis: Wang Hua, Marco Clark, Francis Gassert, Edwin Schmitt, and Qianwen Xu. The arguments made in this chapter are mine alone and do not represent the consensus of the group.

1 Graeber (2001) mentions a third type of value that is less relevant to my analysis: *value* in the linguistic sense, by which meaningful differences, such as between words or morphemes, are constructed.
2 The other corporations hold the rights to electricity distribution on various regional power grids. The predecessor of the State Power Corporation of China was the Ministry of Electrical Power.
3 Five dams are planned for the downstream reach of the Nu/Salween in Myanmar, including the controversial Tasang Dam, which, at more than 185 meters, would be the tallest dam in mainland Southeast Asia. Initial financing has been provided by Japanese government agencies and corporations, with technical and engineering assistance from Sinohydro Corporation and the China Southern Power Grid Company. The precise portion of electricity output that would be sent to China is currently unknown (International Rivers 2009).
4 By contrast, research on the Lancang River (Upper Mekong), the next major watershed to the east of the Nu, suggests that compensation policies have been more reliably implemented in recent years. Villagers currently being resettled for the Nuozhadu Dam, for example, are receiving roughly ten times the compensation sums of villagers who were moved to make way for the Manwan Dam two decades ago, even adjusting for inflation (see Galipeau, Ingman, and Tilt 2013; Tilt 2015).
5 The corollary in urban areas is termed *chaiqian* (literally "demolishing and relocating") and is used to describe land requisition for urban redevelopment. It is one of the most controversial social issues in China today, involving thousands of protests each year, some of which turn violent. The controversy surrounding *chaiqian* stems in part from the ambiguity of current property law in China and the competing interests of government agencies, rural collectives, citizens, and private development companies (see Zhang 2010). In some cases, villagers give up their land rights in exchange for a residential household registration (*jumin hukou*) that allows them to live in urban centers. Meanwhile, private development companies and local governments, who "stir-fry the land" (*chao di*) by converting agricultural fields to residential or commercial property, can reap huge profits.

6 The NGO International Rivers maintains an up-to-date spreadsheet containing facts and figures on dam development in China, a useful resource on a topic that changes regularly and quickly (see International Rivers 2013).
7 There is growing evidence, though no solid consensus among experts, that large reservoirs can exert such pressure on the earth's crust that they cause seismic shifting along fault lines, a phenomenon called "reservoir-induced seismicity."

5

SLAUGHTER RENUNCIATION IN TIBETAN PASTORAL AREAS

Buddhism, Neoliberalism, and the Ironies of Alternative Development

Kabzung and Emily T. Yeh

The hegemonic status of development in China today is reflected in ubiquitous billboards proclaiming the need for a "scientific development view," "green development concept," "fast and speedy development," and "Great Leap Forward development," the last of which is a phrase surely haunted by both the ghosts of capital and shades of Mao. Pressure to develop has intensified significantly in western China since the launching of the Open Up the West campaign in 2000. In high-altitude pastoral areas of the eastern Tibetan Plateau, where animal husbandry is the primary form of livelihood, local governments have sought development through the livestock industry by setting up economic parks for yak products and cultivating a "vision of commodity production" among Tibetans, encouraging Tibetan herders to become market-oriented, rational economic actors who strive to maximize their livestock off-take rate. That is, state protocols to establish a "postsocialist" order to raise standards of living have been predicated on the creation of neoliberal Tibetan subjects. As a result of these efforts, Tibetan herders have, over the past two decades, been selling ever-larger numbers of their livestock to Hui and Han middlemen, who transport hundreds of thousands of yaks to urban markets each year.

However, this increased slaughter has become an issue of great concern to Tibetan religious leaders, particularly of the Nyingma school. According to Tibetan Buddhist principles, killing is one of the most serious sins (*sdig pa*) that can be committed, and it should therefore be avoided at all costs. Slaughter for the market, rather than for personal consumption, is particularly problematic because of the suffering of the livestock on their way to slaughterhouses. Despite ongoing restrictions on Tibetan religious practice and cultural expression, Tibetan Buddhist elites such as Khenpos Jigme Phuntsok and Tsultrim Lodroe[1] continue to wield tremendous social influence and moral authority, which they have used to initiate a slaughter renunciation movement across the eastern Tibetan Plateau. Through religious teachings, they have sought to persuade Tibetan herders to take oaths to stop selling their livestock for slaughter—the opposite of what state officials say herders must do to achieve the goal of becoming developed. Despite the fact that the slaughter renunciation movement would reduce their income, which is ever more important in the context of the competitive pressures of secular economic development, many Tibetan pastoralists have responded to the appeals of these Buddhist elites by taking oaths to stop selling their yaks for periods of time ranging from three years to the rest of their lives. The movement originated in Serthar County, Sichuan, around 2002, and has since spread to many pastoral areas of the Tibetan Plateau.

Beyond the specters of Marx that haunt contemporary China, then, there are also Buddhist critiques of secular capitalist development based on principles of karma, and the privileged status of *all* sentient beings, not just humans. The Buddhist notion of karma, or the law of cause and effect (*las rgyu 'bras*), along with the associated principle of fortune, or *sonam* (*bsod nams*), is quite literally one of "haunting," an abiding presence of the past not only in the current moment but also in the future. *Sonam* refers "not only to a past which determined its accumulation or the present of its deployment but to the future of the consequences that a non-virtuous use may bring in this or the next life" (da Col 2007: 225). With origins in the more distant past than the ghosts of Mao, Marx, or indeed the rise of the capitalist mode of production, the principles of karma and *sonam* concern themselves with a much longer horizon, a continuum of an uncountable number of lives spanning eons of universal time, against which the accumulation of material wealth or indeed of social justice in this lifetime come to look insignificant. Their spectral presence produces, through their interpretation by Buddhist authorities, a new set of protocols for Tibetans, which we suggest both critique and reinforce other dominant policies, practices, and procedures of contemporary China.

Based on ethnographic fieldwork conducted by one of us (Kabzung) in Rakhor Village, Hongyuan County, Aba Prefecture, Sichuan, over a period of eleven months in 2010, we argue that the slaughter renunciation movement is the product of the encounter between the protocols of secular capitalist development and those of Tibetan Buddhism, in which contemporary Tibetan religious authorities have sought ways to reverse certain social and cultural changes brought by hegemonic development projects in pastoral areas of Tibet. Against a secular neoliberal subjectivity, Tibetan khenpos seek through a new Buddhist modernist movement to enact a moral correction upon Tibetan herders and to shape an alternative Tibetan Buddhist subjectivity for the twenty-first century. At the same time, however, the khenpos' critiques posit pastoralism, long a key element of Tibetan culture, as a fundamentally sinful activity and propose that, to avoid livestock slaughter, herders should engage more deeply with state education and consider becoming entrepreneurs in the market economy. In this way, we argue, their oppositional efforts ironically converge with those of state officials in guiding Tibetan herders to become neoliberal subjects. Ultimately, their efforts to articulate alternative forms of development and subjectivity for Tibetans along religious-ethical lines have led to new debates about the fate of Tibetan culture in development, and they have also allowed for the deepening of processes of capitalist expansion. That is, in navigating a rejection of the antireligious politics of communism under Mao, the purely secular logics of Marxian critique, and the competitive pursuit of wealth through livestock accumulation suggested by a new capitalist order, they have created a new Buddhist, but nevertheless still neoliberal, way of being Tibetan in China today.

Yak-Based Development and Secular Neoliberalism

The process of neoliberalization that began with Reform and Opening Up in the early 1980s unfolded in Tibetan pastoral areas first with livestock privatization, followed in the 1990s by the grassland use rights contract system, which is based on the "tragedy of the commons" assumption that common property inevitably leads to degradation, as well as equally flawed but deeply held assumptions among Chinese scientists and policy makers that Tibetan herders are lazy and "irrational." Numerous studies have shown that reports of severe, ubiquitous grassland degradation across the Tibetan Plateau are based on flimsy data that do not stand up to scrutiny and that the localized degradation that does exist is likely attributable to other interacting factors, including vegetation trampling and changed grazing patterns caused by reduced mobility and enclosure, altered herding strategies due to broader policy and political-economic transformations, and climate change (see Banks 2003; Bauer and Nyima 2010; Cao et al. 2013; Harris 2010; Ho 2000; Williams 1996, 2002; Yan et al. 2005; Yeh and Gaerrang 2011; Yundannima 2012). Some of these studies also suggest that the grassland use rights contract system has itself contributed as much if not more to degradation than has overgrazing per se. Yet despite increasing evidence, these neoliberal and agricultural (rather than pastoral) logics and policies persist in practice (Yundannima 2012).

The deepening of market processes was extended by the 2000 Open Up the West campaign, which has been operationalized in specific projects, including infrastructure improvements that link pastoral areas with larger regional markets, housing projects, and yak-based economic parks. Their overall strategy is to transform the historically extensive pastoral system into a more intensive production system by introducing science and technology, commercializing livestock production, and corporatizing the yak industry.

As one of the most productive pastoral areas on the Tibetan Plateau, Aba Prefecture of Sichuan Province has promoted the yak economy as a key development strategy since reforms began in the early 1980s. In Hongyuan County, the two largest state-owned yak milk and meat factories have been privatized and contracted to outside companies in an effort to increase production. The government has also supported local entrepreneurs in the industry, bringing the number of local slaughterhouses to fifteen. A number of companies now produce yak jerky as a "green" product of the Tibetan Plateau to sell in cities and tourist sites. At the same time, new institutes have been established to research yak breeding and genetics, predicated on the idea that trade in yaks will bring great economic benefits. These efforts to promote yak production as a vehicle for development include the establishment of an Aba Prefecture Yak Affairs Office and the declaration of the prefecture's five pastoral counties as a yak economic park. Another effort is the professionalization of Tibetan herders in the yak production system by encouraging them to establish their own professional cooperatives.

These varied efforts to promote the yak economy have been realized and circulated through significant improvement in transportation and communication infrastructure, including a newly constructed road from Hongyuan to Chengdu, the capital of Sichuan Province, as well as mobile phone service and the Internet. The improvement in infrastructure has brought more tourists to the area and accelerated integration with larger markets, thereby increasing demand for yak products and raising prices, and thus giving herders incentives to sell more yaks. At the same time, new regulations driven by concerns about overgrazing that stipulate the maximum number of livestock that herders are allowed to keep, based on officials' estimates of grassland carrying capacity, have also encouraged herders to increase their off-take rate. As a result of all of these factors, Tibetan herders have since the 1990s been selling increasing numbers of yaks to Han and Hui traders. One clear indication of this increased slaughter rate is the fact that Tibetan herders now sell male yaks as young as three years old, whereas before the 1980s they were kept until at least the age of six. Out of an estimated four hundred thousand total livestock in Hongyuan County, more than twenty thousand are now sold to meat markets annually.

This increase in the rate of yak sales for slaughter has gradually altered herders' relationships with their yaks. Their awareness of a karmic relationship with yaks, bound by the cycle of reincarnation in samsara, has become weaker and has been partially replaced by a market-oriented, material, secular relationship in which the very process of selling decreases their fear of the accumulation of sins and other negative results of slaughtering as a misconduct. Rather than a view of yaks as fellow sentient beings, the neoliberal economic context inculcates an alternative cultural view of yaks as commercial products and economic resources. With the replacement of yaks by motorized vehicles for transportation, there are no longer economic rationales for keeping more than a few male yaks once they are large enough to sell for meat. From an economic perspective, keeping them beyond this age simply takes grassland resources away from female yaks, which can bring in income from dairy products as well as further reproduction.

Thus, as a fundamental state strategy since the 1980s, deepened marketization has not only altered the material landscape but also deeply influenced Tibetan herding practices, production systems, desires, and values. In this chapter we focus on state protocols of deepening marketization, while recognizing that the idea of a reified singular state is itself an effect of power, and the site of struggle, contradiction, and multiple hauntings. If "development as the first principle" is the legacy of Deng Xiaoping, then "the power of the pen and the power of the gun" (particularly the gun) is the legacy of Mao that gives development such force. Ongoing arguments between those who champion the current authoritarian capitalist path and those who articulate a New Left Marxian critique are not resolved—but this is not the focus of our discussion here, as the specters of Mao, Marx, and the logic of capital are glued together however fragilely by nationalism vis-à-vis the project of developing Tibet.

Development is not merely a reshaping of economic activities, but also a project of government that shapes its subjects. Building on Nikolas Rose's (1999) conceptualization of neoliberalism as a technology of rule that relies on the power of freedom in the market, Aihwa Ong and Li Zhang (2008) see various Chinese privatization programs as calculative techniques that govern Chinese citizens by providing individuals with ownership of property and optimization of their abilities in the market. This technique of self-governance has placed at a greater distance the previous state presence in everyday aspects of people's lives, leading to a form of neoliberalism that they call "socialism from afar." Lisa Rofel (2007) has argued that Chinese neoliberalism is about the cultivation of particular types of desires. Building on both, we argue that Chinese neoliberal arrangements work to produce not just market-oriented actors but specifically secular market actors.

Though market reforms brought with them an opening for religious practice, they did so only for religion that is officially recognized and thus controlled by the state. In this protocol of the proper management of religion, "religion" is distinguished from "superstition" (*mixin*) and "evil cults" (*xiejiao*), categories that are criminalized and used ultimately to bring religion under state management (Ashiwa and Wank 2009; Dunch 2008; Makley 2014; Yang 2008). At the same time, religious sites have been transformed into tourism assets, arguably also a process of secularization that works to contain and manage religion (Gaerrang 2012). In the Tibetan political context, religion is particularly fraught given the purported association through the Dalai Lama between Tibetan Buddhist authority and a tendency toward separatism. Thus, secularization is assumed to contribute to Tibetan loyalty to the PRC and regional stability.

Development is a project of rule (Li 2007) that works to produce governable subjects, and in contemporary China these subjects are secular and materialistic. In Tibetan pastoral areas, desires for worldly comfort and enjoyment are strongly encouraged by state authorities, while spiritual interests are routinely labeled, through the careful policing of the boundaries of officially accepted "religion," as superstitious and dissident and thus are discouraged or criminalized. Government officials criticize herders' unwillingness to slaughter yaks for religious reasons, claiming that Tibetan herders are irrational and backward for keeping too many yaks rather than selling them to improve their living conditions and educate their children. Instead, the officials suggest, to become developed, the herders must shed these limiting views and aspire to material improvement.

This cultivation of neoliberal desiring subjects is also illustrated in the most spectacular project in pastoral areas of Sichuan Province in recent years, the Housing Project for Herders. Launched in 2009, the four-year program was to provide ninety thousand households with improved or new houses. Of the costs for this program, 20 percent are subsidized by the state, 30 percent are provided by state loans, and the remaining 50 percent by herders themselves. The goal is to ensure that all pastoralist households have their own houses, while also transferring the site of their labor from pastures to secondary and tertiary industries, encouraging herders to settle in towns and begin making a living there. At the same time, the project encourages the remaining herders to improve and modernize their livestock production, increasing their off-take rate. Prefectural and county governments have promoted these goals through interventions including building shelters for livestock and grass storage, encouraging the cultivation of fodder, and developing and disseminating improved yak-breeding technologies and skills.

In Rakhor Village, two previously poor households were designated as model households because of their material improvements and aspirations. Without enough yaks to make a good living in the past, both households settled in town and began working part-time in slaughterhouses. With this income and participation in the Housing Project for Herders, they were able to elaborately furnish and decorate their new houses. In the eyes of other herders, these two households were still considered very poor. However, state authorities viewed them as model citizens who had learned how to improve their living condition and enjoy material comfort while overcoming their religious objections to working in a slaughterhouse. As a result, they have been promoted as models for other Tibetans who have hundreds of yaks but do not live in well-furnished houses and who are thus viewed as lacking the desires and values appropriate for Chinese citizens in the twenty-first century.

Teachings of the Slaughter Renunciation Movement

The unprecedented increase in the slaughter rate of livestock reflects the force of the secular market economy in transforming Tibetan cultural norms and values; the slaughter renunciation movement seeks to intervene in these social and cultural changes. The movement began with teachings by Khenpo Jigme Phuntsok (1933–2004), who established the Serthar Buddhist Institute in 1980 and was the most influential lama in the Nyingma school of Tibetan Buddhism in contemporary Tibet. Known locally as Serthar Gar or Larung Gar, the institute became one of the largest and most influential centers of Tibetan Buddhism in the world, with some ten thousand monks, nuns, and lay disciples at its peak in 2000. It also fostered many khenpos and monks who have gone on to lead other monasteries throughout the region. It became a center for a number of Buddhist modernist movements, including not only slaughter renunciation, but also vegetarianism, the renunciation of wearing animal pelts, practicing the ten virtues, and the purification of Tibetan spoken language (Gaerrang 2012; Gayley 2011; Germano 1998; Phuntso 2004; Yeh 2012, 2013).

Through the 1990s, Khenpo Jigme Phuntsok witnessed the increasing slaughter rate of livestock and became concerned about their suffering. One of his religious teachings in Larung Gar in 2000 marked the beginning of the slaughter renunciation movement. Below are excerpts from the speech he delivered as he sat on a throne on a high slope in the narrow valley, while thousands of laypeople, monks, and nuns listened below:

> During transportation to some cities near pastoral areas, the animals suffer as much as if they had been killed many times over. For instance, after many days on a truck without being able to drink a drop of water, the deathly thirsty animals jump out of running trucks when they see a river near the road, and in the process break their legs. . . . Others are so hungry that they gnaw on each other's hair. . . .
>
> After many days of waiting for the slaughter, the animals are finally brought to where their throats will be sliced with butchers' knives . . . the hot blood from the heart is ejected from their throat, pouring everywhere. . . .
>
> Inside some local slaughterhouses, the scene is similar to what we imagine as the city of death, full of terrifying noises . . . the sound from sliced throats, the sound of running blood. . . . Some yaks are too frightened to go into the building, so the butchers gouge their eyes out so they are easier to move in. . . . The yaks struggle with no hope of survival, but before they are dead, their skins are stripped off and their innards removed.

In addition to these horrifying descriptions of suffering during transport to and inside slaughterhouses, Khenpo Jigme Phuntsok emphasized that all sentient beings share a desire to live and a fear of death, and consequently that humans should not kill other beings for their own needs. "Killing involves the most serious sin. The action of slaughtering causes a person to go to hell for long term suffering, and even after they take rebirth elsewhere, the killer (and seller of livestock) will have a shortened life through many reincarnations. Killing sentient beings causes rebirth in miserable places." In addition, drawing on another Buddhist notion that all sentient beings take so many reincarnations that all may have been, or will be, the mothers of all others in some lifetime, he stated that killing livestock is equivalent to killing one's own mother. How, he asked, could one have the heart to kill one's own mother? Moreover, he emphasized that the accumulation of merit through actions such as chanting mani prayers and building stupas would be negated if accompanied by actions of harming others.

Khenpo Jigme Phuntsok did not confine his teaching to religious principles. He also claimed that "from a worldly perspective, businesses involved with killing will never bring economic improvement. I personally have never seen any middlemen, who trade between Tibetan pastoral areas and nearby Chinese cities, make money from this sinful business. I have seen many slaughterhouses that have killed millions of livestock for many years go bankrupt, one after another." Sinful activities have, he claimed, never brought any economic prosperity to Tibet. Thus, rather than bringing true economic development, the recent increase in the slaughter rate had instead transformed the Tibetan landscape from a place where Buddhism prospers to a demonic place full of hurting and killing. Consequently, he requested that Tibetan herders reduce, or completely halt, their sale of livestock to commercial markets. Following this, Khenpo Jigme Phuntsok and his disciples, particularly Khenpo Tsultrim Lodroe, began to suggest that herders take oaths not to sell livestock for slaughter in the marketplace, for periods of time ranging from three years to the rest of their lives. They also periodically use donations from disciples from eastern China to purchase livestock headed to the slaughterhouse for *tshe thar*, a practice in which the livestock are ritually liberated from slaughter and allowed to graze until they die a natural death.[2]

We argue that this slaughter renunciation movement is the product of the encounter between Tibetan Buddhism and secular neoliberal development. Buddhist authorities interpret the latter as a process in which herders' attention to other-worldly spiritual goals have been dislocated in particular ways by this-worldly material needs. One herder in Rakhor noted, for example, that these days herders need more and more money because of competition within pastoral society over cars, houses, and interior decoration, as well as the high costs of education and health care. He remarked, "The current society is like a magic force that leads us to buy more and sell more," a statement reminiscent of South American peasants' narratives about making pacts with the devil to raise their standard of living but only receiving unproductive consumer goods

(Taussig 1983). This magical force, the transformation of people's worldview by neoliberal development, compels its subjects to accumulate worldly goods, cultivating morals that are inconsistent with Buddhist ethics.

At stake is also the question of whether yaks should be seen primarily as commercial products or as sentient beings—a question that sheds light on larger issues of what development is and how it should be achieved. Within traditional Tibetan cultural views, yaks were livestock to which Tibetans owed gratitude for providing them with things that they needed to live; they are also sentient beings like all others. This status always posed ethical quandaries for herders, who often hired others to slaughter their livestock and who sought to accumulate merit to offset the sins associated with pastoralism. With economic development, herders' daily participation in processes of exchange, price bargaining, production, investment, and consumption has made slaughtering a sheep or yak for the meat market more a process of production, a transformation of one kind of material into another. In this view, livestock are not primarily sentient beings, but rather profitable capital and disposable products that can be exchanged and slaughtered without consideration of factors other than price and quality. The relationship between herder and yak ends with the act of exchange or slaughter, whereas according to Buddhist philosophy this relationship lasts forever in the cycle of samsara, until one reaches enlightenment. To challenge the growing pervasiveness of the calculative dispositions of the secular view, khenpos have argued that just as humans should not sell, kill, torture, or sell the meat of other human beings, so the same is true of the livestock they raise.

More fundamentally, these teachings can be seen as a challenge to the dominant Western ontology of modern technology, in which nonhumans are perceived as mere resources or "standing reserves" for human exploitation and modification (Heidegger [1954] 2008). This orientation of "enframing" the world of beings as a mass of undifferentiated potential resources creates a dualistic relationship between humans and others, and subject and object, that stands in contrast to a Tibetan Buddhist worldview based on cause and effect, or karma, in which life and spirit persist through reincarnation. Human beings are not born to conquer the world and transform its resources, but rather to live harmoniously with them and escape from the bitterness of samsara.

In these ways, the teachings of Khenpos Jigme Phuntsok, Tsultrim Lodroe, and others about the problems of slaughter and their request that herders take oaths to stop are modernist Buddhist attempts at ethical reform for Tibetan herders in the context of a secular neoliberal sociocultural transformation.[3]

These and accompanying teachings do not seek to simply preserve Tibetan culture and Buddhist understanding and practice, but rather incorporate new elements, strategies, and methods addressed specifically at being Tibetan in the world today (see Gaerrang 2012; Gayley 2011). As such, they are a form of engaged Buddhism, a hallmark of Buddhist modernism that reinterprets fundamental teachings in the context of contemporary social needs in specific historical and cultural contexts. The slaughter renunciation movement is not a passively preserved remnant of traditional Tibetan culture, but rather religious elites' active interventions into the social transformations brought by hegemonic forms of development in order to make it more compatible with their interpretation of Tibetan Buddhist norms and culture. They seek to alter development by reducing its displacements of these norms.

More specifically, through the establishment of ethical guidelines for Tibetan herders, khenpos seek a better alignment of development with a Buddhist perspective by reinforcing herders' correct relationships with other beings. In so doing, they seek to show a path for herders to achieve optimal material and spiritual conditions for this and future lifetimes, rather than focusing on this lifetime alone. Theirs is not an outright rejection of development, an antidevelopment view, but rather an engaged creation of alternatives that changes the time horizon under consideration. Buddhist authorities accept the premise that Tibetans are lagging behind others, and must catch up, but their interpretations for both the reasons for becoming underdeveloped and how to overcome this state of being—by paying more attention to the protocols demanded by Buddhist laws of karma and compassion—differ from those proposed by secular neoliberal ideology.

The Movement in Rakhor

In Hongyuan County, Sichuan Province, local lamas inspired by Khenpo Jigme Phuntsok began to teach about the sinfulness of mass slaughter for the meat market as early as 2003. The movement gained significant momentum at the end of 2006 when khenpos from Rakhor Monastery, who had studied at Larung Gar, and a reincarnated lama from another township invited the influential and charismatic Khenpo Tsultrim Lodroe to come and give teachings in Rakhor Village. Prior to the visit, village leaders and khenpos from Rakhor agreed with Khenpo Tsultrim Lodroe that if he visited, the village leaders would persuade herders to take oaths not to sell their livestock for slaughter. That this happened indicates the translation of religious to secular state power, with khenpos realizing their goals through village leaders who are supposed to represent the state's agenda in the current administrative structure, as well as the overlap between this structure and traditional socio-territorial ("tribal") groups that formerly and still form bases of authority.

During the two-day teaching, Khenpo Tsultrim Lodroe gave several empowerments and taught a number of basic Buddhist principles and practices. However, he spent most of his time articulating his concerns about the increased slaughter rate and urging herders to stop their sinful actions, which he stated would lead not only to negative consequences in future lifetimes, but also misfortune in their current lives. His opening remarks reflect the main tone of his lectures in Rakhor:

> I know herders in these areas very well. I have been rescuing yaks from the slaughterhouses in Chengdu for the past half decade, so I have met some of you. Herders [from this area] have been selling large numbers of livestock to the meat market and using that income to build stupas and Buddhist statues, believing they are accumulating good karma. This is a big mistake. I have never heard that killing is good karma in any way. Buddha never taught such a thing.... If you want to practice Buddhism... the first thing you should do is to stop selling livestock to slaughterhouses.... I have traveled many times in this area, but I was very upset to see that herders' belief in Buddhism has become progressively weaker.... It does not have to be this way.

On the second afternoon of his teachings, Khenpo Tsultrim Lodroe asked herders to take oaths for various commitments. A majority took lay vows, which include oaths to refrain from drinking alcohol, committing murder, cheating, stealing, and engaging in immoral sexual conduct such as prostitution and adultery. Many also took vows not to gamble or smoke, which are not part of the standard Buddhist lay vows. As part of a particular teaching, many committed to seven days of meditating on the Four Practices every year for the rest of their lives, in order that they might be born in the Sukhavati paradise of the Amitabha Buddha (*'od dpog med kyi rgyud bzhi bo'i*). In addition, all of the monks, some elderly people, and some young women took vows not to eat meat for three years or longer, and many herders vowed not to eat meat on the most holy days of the year and on the two most holy days of each month.[4] The most difficult vow, which the whole village took, was to not sell their yaks to market for three years; some also took an oath not to slaughter livestock for their own consumption. The majority of households were able to keep their oaths for the three-year period, as was also the case in other villages in the county where the movement took hold.

Because yaks are the main source of income for the majority of herders in Hongyuan, the movement had a significant economic impact. Among the 185 households of the village, approximately 30 are wealthy and own between 150 and 500 yaks; 70 are considered medium-wealthy and own between 50 and 150 yaks; 45 households own between 15 and 50 yaks, while the remaining 40 or so households are viewed as very poor and own fewer than 10 yaks each—with most of the latter having already moved to town. Given the relative market prices of milk and cheese versus livestock, as well as typical herd composition, households on average stood to lose about 45 percent of their income by refraining from selling yaks to the market (see Gaerrang 2012 for more details). However, herders interpreted and responded to this reduction in quite different ways, which affected whether or not they chose to take a second oath after the first one was up at the end of 2009.[5]

Materially, the oath had the least effect on herders who did not have many yaks in the first place. Among the forty-odd households who had already settled in the county town or village settlement and were working in business and seasonal construction jobs, most took oaths for a second period. By contrast, many of the wealthiest herders did not take up the oath for the second period. For them, as for the poorest of herders, the economic losses were relatively insignificant; many of them had savings in banks and enough income from dairy products to cover their expenditures without selling livestock for meat. Instead, for them, the problem was that they simply did not have enough pasture to sustain the increased number of yaks.

The hardest hit by the reduction in income from the oath were those families who were still herding but had relatively few female yaks. For example, a household with five members and forty female yaks out of one hundred would not need to look for alternative sources of income, given that selling dairy products from their female yaks would be sufficient for household needs. By contrast, a similar household owning fewer than twenty female yaks would need to search for sources of supplemental income, such as from operating small businesses or from temporary wage labor. However, such income alternatives are very limited for Tibetan herders, given their lack of Chinese-language fluency and other employment skills.

The movement also changed how herders made use of yaks. Some households sold two- or three-year-old yaks to wealthier families for the latter to expand their herd size; they also started to sell adult male yaks to families that promised not to sell them to slaughterhouses until the end of the oath period. (Most families did not sell their female yaks in this way because of their use in producing milk.) The monastery established a rule whereby anyone who wanted to sell their livestock had to register their livestock with the monastery, to ensure that the livestock buyer would also take a three-year oath. Male livestock sold in this way brought in only half as much income as did a sale to the market, though, so these new strategies only partially mitigated the oath's impact on household income.

Beyond all of these material factors, the loss of income was also interpreted differently by herders of similar or identical means. That is, general trends aside, herders facing similar pasture and income constraints in all wealth categories nevertheless made different decisions regarding participation in a second oath. This suggests that the contestation between secular neoliberal development and Buddhist authority has produced a range of subject positions among herders, even within a small village. Some herders stated that there was no real difference between selling and not selling their yaks for slaughter in terms of the quality of their lives and livelihoods. Some said this was due to the positive karma accumulated, while others explained that the money from yak sales was spent very quickly, but that the lower income from activities such as selling milk or collecting herbs is "higher-quality" income that lasts longer. That is, monetary income from sinful activities is less useful and runs out more quickly than income from other activities, so there was no net negative impact.

Other herders stated that the movement had an impact on their spending, which prevented an overall decline in livelihood. For example, one explained that, in the past, when they sold their livestock they sometimes still faced a serious shortage of cash in the spring, because livestock do not produce milk or are too weak to sell during that time. After taking the oath, he said, they were no worse off than before in terms of cash shortage during this period of time. This was because they formerly spent more on unnecessary things such as gambling, alcohol, restaurants, or other forms of entertainment—forms of expenditure and consumption that constitute an important protocol of contemporary China. They changed these patterns because Khenpo Tsultrim Lodroe had expressed considerable concern about Tibetans wasting their money on frivolous things. With less expenditure, their lost income was not very significant.

Some herders found that their income reduction was in fact significant, but they interpreted the reduced income as being compensated for by spiritual gain. They explained that when they sold many yaks, they made lots of money but always felt guilty and fearful about the associated sins. However, when they did not sell yaks, they felt they were accumulating good karma for their current and future lives, which is more important than accumulating money. All of these experiences and interpretations of the effects of slaughter renunciation suggest something quite different from the ideal *homo economicus* driven by a rational calculation of economic gain. The spirit of critique that drives this alternative mode of being is not Marxist but rather a Buddhist modernist interpretation, which, it seems, is producing new Tibetan subject positions that, as the introduction to this volume puts it, "potentially stand at cross-purposes with the nominal objectives of the national political economy." But do they really?

Ironies of Alternatives

Integrated with his teachings about the problems of selling livestock for slaughter, Khenpo Tsultrim Lodroe also covers various other topics, relating them all to the broader issue of development. He maintains that Tibetans do need development, but the correct way to achieve this is not the mass slaughter of livestock. Instead, he gives generationally specific advice about alternatives— some of which are the same in his teachings across pastoral areas, while others are location-specific. Four common suggestions cut across geographical sites. First, he argues that Tibetan pastoralists must send their children to (state) schools because their only future path lies in acquiring both traditional and new knowledge. Second, he encourages adult herders to make a living by working as wage laborers in towns, in construction, in the service industry, or in any other industry not related to killing or harming. Third, he suggests that herders learn new technical skills and crafts. Fourth, he encourages herders to make money by engaging in honest business that does not involve killing or harming animals. For example, in a teaching near Qinghai Lake, he was very supportive of former pastoralists who had turned to making a living through the tourism business rather than by herding livestock. Similarly, in a teaching in Golog Prefecture, his first suggestion was for herders to make a living by collecting caterpillar fungus and other medicinal herbs rather than selling livestock. In addition, he also suggests selling dairy products and wool.

Ironically, these suggestions and efforts overlap with state efforts to promote development in pastoral areas of the Tibetan Plateau over the last several decades, particularly by encouraging herders to send their children to school, which many were reluctant to do, and to integrate into the market economy. This overlap can be seen by comparing a religious teaching by Khenpo Tsultrim Lodroe with a statement by the Party Secretary of Sichuan Province.

In Rakhor Village, Khenpo Tsultrim Lodroe stated:

> Economically, Tibetans are lagging behind and Tibetans need more economic development, but slaughtering yaks and gambling will never bring prosperity to herders.
>
> ... If we continue what we are doing now, after ten or fifteen years, the future scenario of Tibetan people will be that some Tibetan herders will be living in the mountains herding yaks and others will be farming in the valleys. In the towns and cities, there will be no Tibetans though there will be many Han and Hui. Why is it that there won't be any Tibetans in towns and cities? Because Tibetans do not have the knowledge and skills to live there.

The Party Secretary of Sichuan Province, Liu Qibao, made a similar observation during his visit to the Housing Project for Herders site for Rakhor Village in 2010:

> By providing herders with public infrastructure and service facilities, [the housing project] is helping Tibetan herders to improve their living conditions and become fully integrated into modern civilization....
>
> The permanent houses for herders in settlement sites should be supported by industrialization to make sure that herders are settled and have jobs.... The local government should guide herders to gain market awareness and commodity consciousness to commercialize traditional livestock production.... In order to realize the transfer of labor from pastures, the local government should enhance vocational training programs to improve the *suzhi* [quality] of herders to further help herders working in the secondary industry and service sector.

Their goals differ in that the latter calls for a commercialized and intensified form of livestock production, predicated on slaughter for the market, which the former rejects. However, both offer a vision in which the right and proper sites of modernity and progress for Tibetans are the towns and cities, where many Tibetans should settle by adopting market-oriented practices. Moreover, both call for a dramatic change in the traditional livestock production system and its accompanying pastoral culture and identity.

State authorities argue that traditional herding is both economically inefficient and ecologically destructive and that this is due to Tibetan herders' backward modes of thinking. These are to be changed through policies and projects (or protocols), including the Housing Project for Herders, promotion of compulsory education, vocational skills training, and the current attempt to establish herder cooperatives. State education programs, including the Nine Year Compulsory Education Program, the "9+3" Vocational Program (a program of three years of vocational training after nine years of compulsory education that is targeted at Sichuan Province's Tibetan population), and the provincial "Ten Years' Action on Education Construction Program," are supposed to improve the *suzhi* of the younger generation of herders so that they can become future human capital to be exchanged on the labor market. The state has also stressed the importance of increasing the off-take rate based on an entrepreneurial logic and ranching model. This would reduce the total number of herders from the pastures and funnel them into jobs in town, while professionalizing those who remain.

In a similar way, the slaughter renunciation movement has also put pressure on traditional livestock production, the Tibetan pastoral way of life, and its associated cultural practices and identity, an argument that has been made by Tibetan critics of the slaughter renunciation movement, including radical secularist intellectuals, as well as some monks and villagers, who suggest that the khenpos do not have a good understanding of, or care about, the realities of day-to-day pastoral life.[6] Reinforcing state education policy, khenpos such as Tsultrim Lodroe strongly encourage herders to become educated: "If you do not let your children become educated, they will follow in your footsteps to make a living by slaughtering yaks and gambling. There will be no place for the next generation of Tibetans in town if they are not educated." In this view, it is the transition from uneducated herders to educated townspeople that will allow Tibetans to accumulate more positive karma for their future lives. Thus development, conceptualized in an alternative Buddhist framework in which improvement of the material standards of living in this lifetime are integrated with the ultimate escape from samsara, also requires urbanization of Tibetans. For the state, education of Tibetan herders will lead to greater cultural integration, ecological security, economic development, and social stability. The effect of both of these positions is a hybrid protocol in which the next generation of herders will be settled in towns and fully integrated into neoliberal social arrangements.

The number of young herders who have participated in compulsory education and subsequently gone on to secure stable government jobs (considered the most desirable option) is very small. Nevertheless, children who leave the pastures when they are young tend not to want to return to herding. Given the limited ability of the state apparatus to absorb the unemployed, and the fact that herding is now encouraged by neither the state nor religious leaders, the future path for most young herders is either to become entrepreneurs or to be employed by those with capital in towns and cities.

One local model from the perspective of the slaughter renunciation movement is the case of an exceptionally successful Tibetan businessman who left the grasslands more than a decade ago and started a business selling materials for Tibetan clothes. He started to run a hotel, hiring a herder from Rakhor Village to work as a doorkeeper for the hotel so that the herder could take care of his daughter attending the county primary school. The businessman is doing ideal work in that he does not need to sell livestock for a living. Profits from the running of a hotel, in contrast, are considered to be clean earnings. Both the herder, who is making an effort to change his daughter's life by providing her with an education, and the businessman, who is hiring Tibetans, are Buddhists making a living in ways that do not involve serious sins, the main goal of the slaughter renunciation movement. At the same time, both are participating in the market economy, one with capital and the other with labor power, fulfilling state goals of development through capitalist exchange.

Khenpo Tsultrim Lodroe teaches against excessive attachment to material wealth and cautions that material development is not the way to achieve genuine happiness and peace. Despite this, he and other khenpos have put great emphasis on doing business as an alternative to selling livestock to the meat market. He does this in part by pointing out that Han, Hui, and other ethnicities dominate most economic niches in Tibetan areas:

> Do you know how many restaurants and shops there are in Mewa [Hongyuan County Town]? . . . Who is making and taking that money away? . . . How many are Tibetans? There are no Tibetans. . . . Most are Han Chinese. . . . A similar case is with the butter and cheese that Tibetan herders have collected from their yaks, but Tibetans do not know how to run a business with them. In all of these cases, Tibetans do the hardest work, but make the smallest profit.

In arguing against the labor market segmentation and ethnic inequalities that have come to characterize uneven development in China, to the detriment of Tibetans' relative income, he also works against traditional views of business as dishonest, and wage labor as lowly. He encourages Tibetans to take hourly or daily wage labor jobs as street cleaners, construction workers, guards, and so forth, and to see these as forms of "fair work and clean earnings" rather than as a last resort for those without enough livestock. Similarly, he encourages entrepreneurship by stating at a teaching at Qinghai Lake, for example, that "if people become very wealthy through their hard work and honesty, it is not a problem with respect to Buddhist norms."

The suggestions made by Khenpo Tsultrim Lodroe and other proponents of the slaughter renunciation movement call for a radical break from the pastoral lifestyle, suggesting significant social-cultural transformation for a nomadic culture that has long been not only a key part of material Tibetan culture, but has also played a central symbolic role in the imagination of Tibetan culture writ large. However, this production system and its cultural practices have become problematic both for the state's vision of a developed, prosperous China and Tibetan religious elites' views of Buddhist ethics in the contemporary world. Though driven by different rationales that suggest different immediate strategies for managing pastoralism, the two ironically overlap in a larger pushing of Tibetan herders off their pastures and into wage labor and business.

This convergence has led to herders' departure from self-sufficient livestock production and integration into a more cash-based market economy. This shifts social relations so that settled herders are no longer independent household production units, as they were in the past, but rather enter into new social relationships in which they are either owners of capital who hire others or those who sell their labor power to others. That is, khenpos' religious movements and secular neoliberal forces have worked together to engender social relationships among settled herders that invite a Marxian critique.

Yet this unevenness is normalized by neoliberal market ideology on one hand and masked by religious norms of karma and *sonam* on the other. In the former, the inherently uneven nature of capitalism is masked by an ideology of free competition, in which differential success or failure is not the result of resource access and distribution but rather is rooted in individual factors such as hard work and adequacy of self-responsibility and self-entrepreneurship. According to Buddhist norms, the condition of lagging behind found among a group of people or the poverty of individual households is attributed to the accumulation of negative karma. Those who live in more developed communities, or those who are lamas and other religious leaders, state officials in powerful positions, successful businessmen, and very wealthy herders, are labeled *bsod nams can*, meaning one who has accumulated good merit or fortune due to good karma. Thus here too outcomes are not attributed to structural inequalities of power relationships or resource access in the present world but rather to karma accumulation and morally adequate behavior over many lifetimes. According to Buddhist concepts of remote causes (*ring ba'i rgyu*) and proximate causes (*nye ba'i rgyu*), what poor people need to do is to accumulate merit and work hard for their future fortune. The expression that encapsulates this principle, *tshogs bsags bsod nams*, was common before 1959 but was labeled superstitious and banned during the commune period, given its direct contradiction with the principles of communism. But the expression became common again in the 1980s, with religious freedom coming to the rescue of justifying the inequalities brought by market reforms. Thus the regimes of neoliberal capitalism and Tibetan Buddhist ethics, antithetical in their views on whether livestock should be viewed as commodities for the market or sentient beings in the cycle of samsara, nevertheless come together to produce new, mutually agreeable protocols for Tibetans in contemporary China, ones that both are inflected by the institutional weight of a vestigial Maoism (in the ongoing state control of religion) and invite a neo-Marxian critique.

Conclusion

Jean Comaroff and John Comaroff (1999, 2000) argue that, in the face of spectacular inequalities and speculative wealth, the failure of development to deliver on its promises has led to the emergence of new religious movements. Pentecostalism in Ghana, for example, emphasizes the dangers of globalization, while presenting itself as the only means through which people can navigate it, rather than becoming its victim (Meyer 1998). Indonesia has witnessed the emergence of a new assemblage of Islamic and capitalist ethics in which being a pious Muslim becomes equivalent to "inculcating the ethical dispositions deemed conducive to market success" (Rudnyckyj 2008, 2009: 130).

In Tibetan areas of contemporary China, a different assemblage of capitalist development and religious authority has emerged, in which state protocols that foster neoliberalism have enabled new interpretations of Buddhism that attempt to correct those secular neoliberal tendencies in an articulation of an alternative form of development. State development protocols have set the conditions of possibility for livestock slaughter to be seen by Tibetan religious leaders as an ethical problem demanding a solution. Prior to 1959, livestock slaughter was much more limited, and moreover there were no other livelihood options for Tibetan herders other than pastoralism. During the high Maoist period, religious leaders had no power to assert their moral and ethical influence over Tibetans. Only with China's rapid development brought by economic reform were the conditions of possibility established for the movement: not only Tibetan khenpos' reestablished ability to exert authority, but also urbanization, infrastructure improvement, state policies to educate and settle pastoralists, and the availability of alternative sources of income.

Tibetan Buddhist leaders seek to domesticate the impacts of China's new development protocols, making them less culturally destructive and consistent with (re)interpreted Buddhist norms and standards. This is a selective process in which Tibetan Buddhist elites and herders embrace some elements of the hegemonic development project, giving meaning to them, while seeking to correct or reject others. Even as they reject the state's calls for yak-based capitalist development through an increased slaughter rate and more efficient production as antithetical to being a proper Tibetan Buddhist in contemporary China, religious elites have selectively embraced other capitalist practices and ideals. In rejecting mass slaughter, they also denigrate the cultural practices of herding, suggesting that Tibetans instead become settled in towns, educate their children, work in wage labor, and become entrepreneurial subjects. These practices significantly reinforce development as the spread of uneven capitalist relations, tying Tibetans more deeply into the Chinese polity and regional and national markets, while at the same time being coded with Buddhist norms and meanings. Contemporary Tibetan subjectivities in China today, then, are contested outcomes of competing projects of secular neoliberalism and religious development. What emerges in the slaughter renunciation movement is a kind of Buddhist neoliberalism, a form of being and a set of protocols haunted by not only the ghosts of capital, shades of Mao, and specters of Marx, but also the footprints of Buddha.

Notes

1 Khenpo is the highest degree in the Nyingma school of Tibetan Buddhism. The authority of lamas is based on the fact that they are recognized as reincarnations, while the designation of khenpos and geshes (the highest degree in the Gelugpa school) is based on their achievements in learning. In some cases, as with Khenpo Jigme Phuntsok, who was both a lama and a khenpo, the two overlap.

2 Lamas and monasteries have traditionally received donations from practitioners, and they continue to receive donations today. Although monks and lamas are not allowed to solicit money, this does not deter disciples from offering, which is viewed as meritorious. Increasingly, lamas of stature have large numbers of Chinese disciples, many of whom donate enormous sums of money (see Costello 2008; D. Yu 2012). Historically, pastoralism played a role in the support of monastic institutions, including monks and lamas, but this role is considerably diminished

today, given other sources of wealth. Instead, khenpos at Larung Gar have helped support pastoralists in nearby areas who have taken part in the slaughter renunciation movement.
3 On Buddhist modernism, see Gayley (2011), McMahan (2008), Queen and King (1996), and Swearer (1996).
4 Animal products traditionally make up a significant portion of the Tibetan diet, and vegetarianism was not common in the past, due in part to the difficulty of growing crops in pastoral areas.
5 Despite the fact that the movement directly contradicted state development goals for the livestock industry, it did not face any direct state interference. In contrast to many areas in Kham where the oaths were renewed, many herders in Rakhor did not renew their vows in 2009. A detailed discussion is beyond the scope of this chapter (see Gaerrang 2012).
6 See Gaerrang (2012) for a discussion of these critiques.

6

"YOU'VE GOT TO RELY ON YOURSELF...
AND THE STATE!"

A Structural Chasm in the Chinese Political Moral Order

Biao Xiang

China today is characterized by the coexistence of a relatively high level of stability at a national level and deep instabilities at the local level. On one hand, the ruling position of the Communist Party remains unchallenged in every formal way, and polls consistently report surprisingly high levels of public acceptance of the regime.[1] On the other hand, social tensions run deep in everyday life, especially among laid-off workers, landless peasants, and badly exploited migrants. People's frustration and anger are clear, but a sense of hope is also palpable. How should we make sense of this contradiction? And what does this mean for China's political future?

Academic literature and public opinion have commonly attributed the Chinese regime stability to certain unified and unifying hegemonies, particularly developmentalism and nationalism. Developmentalism justifies the Party's rule by its performance in delivering economic growth. But as Vivienne Shue (2004) has observed, a regime basing its legitimacy on performance alone is intrinsically unstable. Furthermore, the Chinese Communist Party (CCP) has itself vowed to move away from the developmentalism paradigm since the 2003 leadership transition (from Jiang Zemin to Hu Jintao). Nationalism, which is supposed to replace communism as the ideological basis of the regime, is a double-edged sword that can turn against the regime at any time. Nationalism has been appropriated by political forces of very different stances throughout modern Chinese history, and the 2012 outbreak of the anti-Japan movements over the Diaoyu/Senkaku Island disputes once again revealed some of its complexities.

In this chapter, I propose to move away from a focus on unified hegemony to an attention to contradictions and fragmentations in explaining regime stability. I argue that social conflicts in China are defused not only through coercion, nationalism, and developmentalism, but also through a structural chasm between the logic of how tensions emerge and the logic of how they are dealt with. Social tensions in contemporary China are typically conflicts over economic interests arising from the marketization of social life that is driven and complicated by local government. The solution, meanwhile, is often political, especially when the tensions become antagonistic. The tensions are typically addressed by the central state in a top-down, case-specific manner, through party disciplinary measures including personnel reshuffling. Such solutions are based on concerns about relatively long-term stability instead of short-term growth, and they are almost always presented in highly moral and ideological terms. I call it a "structural chasm" because the political solution hardly addresses the root causes of the economic conflicts. The structural chasm provides quick fixes precisely because the logic of problem solving is different from that of problem generating and because the solution lies outside of the domain where problems emerge (the political versus the economic).

Such a structural condition gives rise to a sentiment that is captured by a refrain frequently voiced by my informants in Northeast China: "You've got to rely on yourself . . . and on *guojia*." That is to say, one has to rely on oneself because it is believed that only entrepreneurial and calculative individuals can survive the market-driven development, but one has to rely on *guojia* because the state is regarded as the ultimate solution to problems. It is important to note, however, that *guojia* is a much broader concept than the state, and also encompasses the government, the country, and the nation. *Guojia* represents a totalizing order and an all-embracing framework for making sense of public life. It carries strong normative and moral meanings. People commonly describe corrupt local officials as "worms" (*zhuchong*) or even traitors of *guojia*. They sometimes voluntarily criticize themselves and fellow citizens for not following *guojia* policies fully or for not having contributed enough to *guojia*. This political conception had a long history in imperial China but is decidedly reinforced through the repeated revolutionary mobilizations since the late nineteenth century, especially by life under state socialism after 1949. Seen from this angle, the public acceptance of the party-state is less an agreement on the nature of the regime than it is a reflection of a political epistemology.

Is this structural chasm simply a manifestation of the well-documented local-central relations in Chinese politics? As has been repeatedly pointed out, protests in China are often against predation of the local government and look up to the supposedly benevolent central state for protection (Chen 2004; Shue 2004: 29; So 2007). "Rightful resistance"—resistance against local government over specific issues by drawing on general principles promulgated by the center (O'Brien and Li 2007)—is the paradigmatic characteristic of those protests. Ching Kwan Lee in her analysis of labor protest points to a situation in which "decentralization makes local government responsible for developing a probusiness local political economy, while the same local government agents are called on to implement labor laws promulgated by the central government eager to resolve labor conflicts and to maintain social stability" (2007: 11), which she aptly summarizes as "decentralized legal authoritarianism."

The structural chasm corresponds to the local-central divide in some ways, but it also goes beyond it. This chasm refers primarily to the gap between two opposing rationalities, namely everyday instrumentalism and abstract moralism. Ordinary people are not passive victims who are exploited by the local government and wait to be rescued by the center. Instead, they can be highly opportunistic in dealing with the local government. They are fully aware of local government's developmentalist agendas and rent-seeking intentions, and they are not reluctant to exploit them whenever possible. A well-known example is peasants' rush to plant trees, "plant houses," and "plant toilets" on their land before it is zoned for redevelopment, in order to demand more compensation. It also became common in the late 2000s for people to "blackmail" local governments by threatening to petition the higher authorities, to the point that the leading Chinese sociologist on popular contestation, Yu Jianrong (2010), called for fundamental reform and even abandonment of the petition system. Yu correctly pointed out that using the petition as a way to tighten the center's disciplining over the local level could well exacerbate tensions, because this would encourage people to use moral principles of the central state to pressure the local government for instrumental purposes. Such tensions between the two logics are what the structural chasm is about.

Can this moralism and instrumentalism be merely reflections of people's different positioning in different contexts that follows the same interest-maximizing strategy? "Speaking human languages to humans, speaking devil languages to devils," is, after all, supposedly common wisdom in the Chinese society. I argue that the current situation is more complex than that. First of all, the central and the local governments are regarded as qualitatively different, and the distinction is more than that between good and bad, or even between human and devil. The central government is supposed to be an embodiment of moral principles, while the local governments are self-interested bureaucratic organs. How, then, can we know whether people sincerely believe in the central state's policies and rhetoric when projecting the central state as a moral agent? Or are they simply acting opportunistically? In most cases, people are perhaps both sincere and opportunistic (O'Brien and Li 2007: 23–24). Few believe themselves to be lying when they pronounce these sorts of glorious slogans, though they may have agendas that have little to do with these slogans themselves. Empirically, we may never know for sure, though I *choose* not to reduce actors into pure instrumentalists. I interpret their seemingly contradictory engagements with the state as an expression of their political subjectivity—a subjectivity that embraces instead of escapes contradictions and that seeks solutions by making direct moral demands rather than by following legal procedures or through market negotiations. The people may or may not believe in the state, but they nevertheless try to see hope in it and engage with it.

The concept of a structural chasm aims not only to describe how reality looks, but more importantly to discern internal possibilities for change. Instead of attempting to predict whether the regime will collapse tomorrow, I am more concerned with what possibilities we have in our hands now. I hope the concept of a structural chasm will help shed light on how the Chinese people engage with an ever-changing, uncertain, and deeply problematic but always hopeful reality, as well as serve as a tool for researchers to engage with such engagements.

Ethnography of Incidents

In probing this structural chasm, I propose a methodology centered around an "ethnography of incidents." An incident is not simply a case or an example, what Marshall Sahlins (1985) calls an "empirical form of system" (153), or what Victor Turner (1974) describes as a revelation of existing antistructure momentum. Rather, incidents and events are first of all "evidence of the ongoing dismantling of structures or of attempts to create new ones," and thus they bring to light what Sally Moore (1987) calls "a multiplicity of social contestations and the voicing of competing cultural claims" (729). In other words, an incident is much more than a dramatization of existing social relations or an instantiation of general conditions, and instead it opens spaces where new meanings and possibilities are produced.

This sort of ethnography of incidents can be particularly productive in probing China's contemporary conditions. Mass incidents (*quntixing shijian*) have been a critical part of Chinese public life since the 1990s. Ranging from peaceful demonstrations to violent riots, mostly apolitical and focused on economic interest, the vast majority of these incidents target local governments but are solved through interventions from a higher level of government authorities. The occurrence of these incidents is unpredictable, but their unfolding is logical. Although triggered by accidents, the incidents are constructed by conscious actions and can be highly consequential. China currently finds itself in "times full of incidents and events" (*duoshi zhiqiu*), wherein incidents are probably *the* most effective and common means through which societal forces push for policy changes.

Incidents do not yield changes in and of themselves, but rather they are consequential by becoming public topics. The term *mass incident* appeared 13.83 times per month in 2010 in *The People's Daily*, the flagship newspaper of the CCP, compared to 3.25 in 2001 (Steinhardt 2012: 8), and in cyberspace these incidents are closely followed, analyzed, and debated. As scholars such as Stephen Ellingson, Regina Lawrence, and William Sewell have pointed out, in a wide range of contexts incidents may significantly "affect the power of a particular discourse or frame" (Ellingson 1995: 136) and enable "critics, activists, and other advocates of change . . . to actively shape public discourse" (Lawrence 2001: 95) in ways that may even lead to revolutions (Sewell 2005: 236–244). Incidents in China today constitute important topics that enable people to express views and develop new consciousness (Steinhardt 2012). An ethnography of incidents aims to bring together the logic of how changes are made, the logic of how the changes are presented and reflected on by the public, as well as the logic of how they are analyzed by researchers.

The incident that this chapter focuses on is a relatively minor one. Seventy-one workers in Liaoning Province, in Northeast China, were swindled after they each paid a private recruitment firm nearly RMB 40,000 (USD 5,700) in order to migrate to South Korea to work in 2004. Most of the victims were between twenty and thirty-five years old, came from small cities and countryside towns, and had junior high–school educations. They were enterprising, risk taking, legally savvy, and financially shrewd. Instead of taking the firm to court, the victims took on the local government as their primary target.

They argued that the local government was responsible because its corruption caused the intermediary's wrongdoing, thereby effectively translating a commercial dispute into welfare concerns of a disadvantaged group resulting from government corruption. The victims negotiated with the local government for financial compensation, and at the same time they appealed to the central state on moral grounds for support. The would-be migrants decided to forego legal options and instead to engage with the local and central governments directly, and furthermore to deal with the central and local governments differently, because, of course, they believed that these measures would serve their interest. However, the questions are as follows: What made them believe that it was feasible to negotiate with the local government? Why did they think it was productive to regard the central state as a moral agent? Like the protracted legal cases of the Fijian farmers who fought for their land rights as described by Hirokazu Miyazaki (2004), my informants' prolonged engagements with the state serve as their methods of hope. I discern their political subjectivity from their hopefulness about their political actions. Before delving into the incident itself, it is necessary to sketch out the "marketization" of labor out-migration, which is how the Chinese official language called the reform of the regulation.

The Marketization of Out-Migration

My informants' motto, "You've got to rely on yourself," is particularly true in the case of labor out-migration from Northeast China. Dubbed "the eldest son of the Republic," the northeast had been dominated by heavy and military industries. In the 1990s, and particularly after 1994, however, a large number of large state-owned enterprises went bankrupt, which led to massive layoffs. Once labor aristocracies entitled to cradle-to-grave welfare, the workers now stood in the market as naked individuals with minimal institutional protection. Out-migration was identified by the government as an important means of reemployment. FS, an industrial prefecture where the incident that this chapter focuses on took place, was hit hard by layoffs and has become a major source of migrants.

Labor out-migration is also encouraged by the liberalization in exit control. By 2005 most Chinese citizens were able to obtain passports in as little as five working days, and the Passport Law of 2007 defines the possession of a passport as a basic entitlement. Regulations over labor out-migration were also reformed. Originally defined as state-managed projects of "international labor cooperation," labor out-migration was redefined in 2002 as individuals' "overseas employment." Labor out-migration increased significantly, and by the end of 2011 there were aproximately 800,000 Chinese working overseas on contracts of more than a year, compared to only 58,000 in 1990 (Center for International Exchanges 2005; China's International Contractors' Association 2004; Ministry of Commerce 2012). South Korea, Japan, and Singapore are the three top destinations worldwide.

New policies established in 2002 allow and even encourage private companies to engage in international labor recruitment. As a result, the number of licensed international labor recruitment intermediaries grew from four in the early 1980s to more than three thousand in 2005 (though the figure dropped to just over one thousand in 2012 due to changes in government policies). These companies are expected to solicit overseas employment opportunities and at the same time to help control migration by selecting the right candidates, filling out the right forms, and following the procedures as stipulated by both China and the migrant-receiving states. Only companies with direct international connections, large assets, and sizable professional teams are qualified for the licenses. Most of the companies are in major cities and have little connection with local communities where manual workers are from. They therefore rely on subagents for recruitment, who may in turn outsource the task to smaller entities or individuals at the grass roots. Local governments, and specifically the bureaus of international economy, are keen to promote out-migration and sometimes issue special permits to companies in their localities as recognized recruitment subagents in order to give them more credibility. The complex relations between the multiple intermediaries are, however, often hidden from would-be migrants.

These reforms created an environment in which a would-be migrant feels that one has to rely on the self, as well as that one *can* rely on the self. As long as you pay high-enough fees, multiple intermediaries would work together to find overseas opportunities for you. The involvement of multiple intermediaries, though creating more chances for out-migration, also increases costs and risks, which further reinforces the notion that one must take risks in order to have a chance of success. Agent fees are perceived as a form of investment. The reform is officially referred to as "marketization" also because government is supposed to withdraw from direct management of out-migration. There is hardly any need for would-be migrants to interact with government agencies, as most of the paperwork is dealt with by the intermediaries. For the would-be migrants, government regulations become much easier to navigate, but they have to pay intermediaries at every step along the way. This process is part of the general marketization of social life in China today.

Forty Thousand Yuan for a Fake Visa

Ma Na, a twenty-nine-year-old woman, was one of the many who were exploring this sort of marketized out-migration. Ma was renting a stall in a marketplace in FS to sell medicines when she heard that a local company, Green Metro Business Development Consulting Center, was recruiting workers for jobs in South Korea. Ma checked all the company's available documents and asked around about the manager, Wang Luli, and was reassured by what she found. Green Metro had a special permit issued by the FS Bureau of International Economy for recruiting workers in FS as subagents for fully licensed companies. More important, Wang Luli's brother, Wang Zuli, was the chief secretary of the FS municipal government. (This position is of the same rank as a deputy mayor or half a rank lower, who is in charge of coordinating government functionaries. A chief secretary should not be confused with a party secretary.) Given that a government official cannot vanish overnight, Ma thought, she could always track Wang down through his brother if things were to go wrong.

There were a series of irregularities in how Green Metro dealt with the would-be migrants. When Ma and seventy other applicants paid Green Metro the compulsory initial deposit of RMB 2,000 each, for instance, they paid the money to Wang Zuli's two sons (Wang Luli's nephews), who were officially unrelated to the company. One of them worked at FS Investment Bureau, and the other was an officer for a local prosecution bureau. Furthermore, the applicants were not given receipts from Green Metro, but rather ones stamped by FS Coal Metro Travel Agency. They were told that Coal Metro, a fully state-owned firm, was "the same thing" as Green Metro. It turned out that Coal Metro Travel Agency was owned by FS Tourism Bureau, and Wang Zuli used to be the head of the bureau. The applicants, however, were hardly troubled by these irregularities. The Coal Metro stamp was seen as being a step closer to the government and thus carrying more weight. Wang Luli also brought up his brother at critical moments. When the applicants asked why nothing happened at the end of 2003, despite the fact that the contract had promised to send the workers out by August of that year, Wang pounded the table and shouted, "My brother is the chief secretary! You don't trust me?"

Good news finally arrived just before the 2004 Chinese New Year. The applicants were told that their visas had been issued and they must pay the final installment of RMB 33,000 (USD 4,700) immediately. Ma Na urged Wang to check with the Korean consulate about the visas. Wang shot back that he had checked everything: "Do you think I know nothing about the stuff of government?" The applicants insisted on seeing at least one visa before making a payment. Only then did Wang Luli tell the migrants that he was actually recruiting on behalf of Shenyang Labor Center, a public institute but at the same time a registered company owned by Shenyang Municipal Labor Bureau. Shenyang Labor Center was holding all the passports and visas and would not release them until after receiving the money in full. Finally, Ma proposed to pay her fee of RMB 33,000 to check her visa on behalf of the team. When Ma arrived in Shenyang, Wang Luli suddenly called her and asked her not to go to the labor center, but rather to a district civil affairs bureau to see a Mr. Huang Jie. Ma was perplexed but did not ask much. "They are all government [departments]," she thought. In the civil affairs bureau Ma went through all the rituals of entering a government compound—waiting, filling forms, being checked by the security guard—before finally being escorted in. Ma saw her passport and visa. She sent a message to her fellow applicants, and all the money was paid instantly.

One month later, while the would-be migrants were busy bidding farewell to relatives and buying clothes and medicine,[2] Wang Luli called them to an urgent meeting and told them that all the visas were fake. It turned out that Huang Jie, a former cadre in the Shenyang Labor Bureau, ran his middlemen recruitment business as a representative of a vocational school affiliated to the bureau. He met a parliament member of the Republic of Korea who had promised four hundred jobs in South Korea and had demanded advance pay. Huang Jie paid him RMB 3 million (USD 430,000) to secure the deal. In order to make the deal appear more credible to would-be migrants, Huang approached Shenyang Labor Center as a partner of the project. At the same time Huang Jie brought in Wang Luli—who used to be on the same college basketball team—as a subagent for recruitment in the FS Prefecture, while Huang also recruited from other areas. Wang signed the contract with the labor center. But suddenly the Korean dropped out of sight. Desperate to recoup his money, Huang Jie flew to Hong Kong and bought more than one hundred fake visas in order to collect money from the would-be migrants. Huang persuaded Wang to transfer the migrants' money to him instead of the labor center (as the contract required) by offering a commission of RMB 8,000 (more than USD 1,100) per head—RMB 3,000 more than the center had agreed on in the formal contract. Huang disappeared immediately afterward.

The would-be migrants were particularly enraged by the fact that they had been cheated even though they had been so careful throughout. Ma had kept all the payment receipts, photocopied documents, and recorded important conversations in her diary, from which I was able to reconstruct the facts of the incident. The would-be migrants were constantly vigilant for others' activities that may be "against the law" (Lee 2007). In recounting the incident to me, the victims repeatedly stressed the importance of law. Lin Feng, one of the victims, had sold his grocery shop to pay for the agent fee, and he was now penniless and jobless. He was particularly emotional, and on more than one occasion he broke down in tears when talking to me. He asked again and again: "Which is greater, [political] power or the law?" ("*Quan da haishi fa da*?"), and "Why is the law so pale (*cangbai*) in the face of power?" It appeared to be a strong consensus among all the victims that the underdevelopment of the rule of law was the cause of all their problems. However, despite their legalistic approach and documentary evidence, the victims made every effort to *avoid* legal actions against the commercial recruiters.

No Lawsuit

The victims' first reaction in the wake of the sham was to stage dramatic public demonstrations. Twenty victims climbed to the top of an eight-story building, opposite the FS Public Security Bureau, and threatened to jump, shouting, "We don't want to live anymore!" and "Refund our money!" Firefighters surrounded the building, laid out air beds on the road, and local media rushed over. Finally, a deputy chief secretary of the municipal government arrived at the scene and promised to conduct an investigation. Two weeks later, when the victims heard no news from the government, another twenty victims gathered in front of the municipal government compound to demand to meet the mayor. Finding themselves blocked at the gate, they knelt down in freezing wind for three hours and shouted repeatedly: "We want to see the mayor!"

A number of meetings between government officials and the victims were arranged after that. From the very beginning, the message from the government was quite clear: The victims must take legal action, including suing the Bureau of International Economy by invoking the Administrative Litigation Law. The Office for Letters and Visits, the focal department dealing with complaints, proposed to pay the victims' legal expenses and even assigned a lawyer at a government-run legal assistance center to work with the victims. Legal actions must be swift, they urged. At that time, Huang Jie had just been arrested and would be tried in the provincial court in Shenyang, and the government lawyer warned that if the victims did not file a lawsuit immediately they might not be able to get their share of compensation because Huang Jie's assets could be divided up by victims from other places. The victims were worried about exactly the opposite—they reckoned that if they *did* file a case in the FS court, the compensation from Huang's assets might be transferred collectively to FS instead of being handed out to individuals in Shenyang directly. As Ma Na put it, "If we receive money *in* FS, the [FS] government will say that we have been compensated already and they have done their job!" They feared that a legal case would leave the FS government off the hook. Furthermore they were worried that the government might maneuver the legal process and push them into an endless legal maze.

The victims' concerns turned out to be not without merit. Under the government officials' encouragement and with the assistance of government lawyers, eight victims proceeded to sue the Shenyang Labor Center. The court judged the labor center guilty. According to Lin Feng and others who refused to join the legal action but followed the case closely, FS government and the FS court made Shenyang Labor Center the scapegoat in order to throw the problem out of their jurisdiction. The indictment proved unenforceable. Shenyang Labor Center refused to compensate the claimants on the grounds that they had never received the money. When the victims urged FS Bureau of Public Security to enforce the verdict, the official in charge replied, "The head of the district court [in FS] is equivalent to a branch director of the provincial [labor] bureau [in administrative rank]. I am one level lower, so how can I possibly enforce [the verdict]?"

Lin Feng and others who refused the legal option insisted that the scam was a result of government corruption. "I would have let it go had we been cheated by private agents. There are plenty of people out there cheated by private agents," Ma Na said. "But we paid money to *you* [Wang] because you are connected to the government! Your brother is the chief secretary of the government!" The victims concluded, "We were cheated because we trusted the government." They argued that the State Compensation Law should be invoked, which stipulates a right to compensation for the damage caused by the unlawful exercise of authority of state organs and their staff. When I asked what corruption was involved, exactly, Ma stared at me: "His brother is the chief secretary of the government! There is *of course* corruption."

The victims soon collected the information they wanted. They found out that it took Wang Luli a letter of less than two hundred characters and two days to get approvals from three government bodies to set up Green Metro. Based on their investigation, the victims identified eight persons and institutions who should be responsible for the Green Metro scam, including Wang Luli's brother Wang Zuli, Wang Zuli's two sons (who had helped collect the money), the legal representative of Coal Metro Travel Agency (for "illegally endorsing Green Metro's business activities"), the Bureau of International Economy for issuing the permit and for its internal decision to reward Green Metro as a model enterprise in FS for its economic performance just before the scam was exposed, the management committee of the economic development zone where Green Metro was located, and the tax bureau for "illegally issuing Tax Registration Certificates with two-month validity." Also on the list was the chairman of FS Federation of Trade Unions. The chairman took a five-day tour, arranged by Coal Metro Travel Agency, to South Korea in April 2002 and Wang Luli had acted as the financial guarantor for his visa application. Thus, according to victims, "there must be something" between him and Wang. Shenyang Labor Center was conspicuously absent from the list. The victims wanted to move all responsibilities to the FS municipal government.

The victims also came up with specific instructions as for what exactly the local government should do. The government should, they proposed, first of all force Wang Luli to compensate the victims and also refund the expenses that they had incurred during the campaign. If Wang's assets were insufficient, the FS Bureau of International Economy, the special economic zone, and Coal Metro should fill the gap. The government rejected this proposal. After Huang Jie's trial in Shenyang, each victim was compensated RMB 14,000 (USD 2,000). The victims developed a new proposal for how their remaining loss should be covered: Wang Luli should pay RMB 200,000 by selling his property; the municipal government, RMB 200,000; and the Bureau of International Economy, RMB 100,000 (respectively, figures of USD 32,000 and 16,000). The government turned this proposal down as well.

Wang Luli was not only exempted from the main targets, but was also turned into a campaign partner of the victims. The victims believed that inside information and close connections with government were crucial for their campaign, and Wang had both. "He knows people [in the government], he can run around, find ways to get more money back," said Ma Na. Before Huang Jie was arrested, Wang visited Beijing three times and went to Shenyang weekly to urge relevant departments to hunt Huang down. After Huang was arrested, the victims pressured Wang to monitor the trial in order to make sure that the FS victims would get their fair share from Huang's assets, as compared to the victims from other areas. Furthermore, they asked, what was the point of sending Wang to jail if the victims couldn't recoup their losses? Indeed, the victims pressured the local government for administrative solutions, rather than pursuing lawsuits, precisely because they believed that the local government had the capacity to pay the money, and administrative interventions could materialize it swiftly. Formal justice and procedural integrity were of only secondary concern.

Rolling-Pork Bargaining

While the victims tried to corner local officials, the officials hid themselves behind bureaucratic procedures and legal technicalities. Ma Na called the officers at the Office for Letters and Visits "rolling pork," a northeastern expression for a kind of pork that always rolls to a side of the knife when one tries to cut it. Here is one exchange with a deputy director from the office, recorded in Ma's diary, when the victims had demanded a clear decision from the government:

> OFFICIAL: We have to handle things according to procedures. We cannot make you any promises right now. This is not children playing games or selling goods in the marketplace.
> VICTIMS: When will the procedure go through?
> OFFICIAL: We will let you know.

Sometimes the officials' explanations were so technical that the victims could hardly understand them. The victims made at least fourteen visits to various government departments over nearly two years, and nearly every time they were received by different officers. They therefore had to tell the story again from the beginning each time, only to get similar vague answers in the end. A government official in FS (not involved in this case) agreed with me that it could be a deliberate strategy for government departments to ask different officers to receive the same petitioners every time in order to slow things down. Individual officers were also reluctant to see the same visitors again and again. "What can he/she say [if the visitor asks] why nothing has happened since last time?" Colleagues in the same office sometimes took it as a duty to meet the same visitors in turn.

As frustrating as they were, the officials never appeared antagonistic. All the officials agreed that the victims deserved compensation, though everyone stressed that it was up to the court to decide who should offer the compensations, and how. The victims had no problem walking into a government department, presenting the case, and demanding to check documents. The Office for Letters and Visits emphasized its "neutrality," rhetorically positioning itself as an intermediary between the government and the victims. Ma Na noted in her diary that officials would say things such as "I hope that we will maintain a good relationship"; "[We should] keep in close touch, exchange information and views all the time"; and "[We] inform each other . . . to avoid the deterioration of the situation, to ensure no death or injury."

As a counterstrategy against the rolling pork, the victims suggested that a top official, the mayor or deputy mayor, "stand up" to "*bao an*"—literally embracing the case in its entirety—and take the full responsibility for settling the case. If the mayor was willing to do so, the victims would promise not to visit the government offices and would leave the officials in peace, but this was ruled out by the Office for Letters and Visits.

Go to *Guojia*!

Five representatives of the seventy-one victims went to Beijing when the *bao an* proposal was rejected. The victims had already phoned the Ministry of Public Security in Beijing to report the case and seek advice in the wake of the scam. During the visit they called on the Ministry of Commerce (which is in charge of local bureaus of international economy), the Ministry of Public Security, and the State Council Office for Letters and Visits. The victims presented petition letters along with detailed written accounts of the dispute and especially of the wrongdoings of the local government agencies. As most petitioners in China do, they wrote the letters in a highly moralistic and ideological tone. CCP slogans at the time were enthusiastically adopted, including "three represents," "harmonious society," "outlook of scientific development," and "disadvantaged groups." The letter accused the local governments for failing to implement the central state's policies and to uphold the "central spirit" (*zhongyang jingshen*). The scam itself became the background instead of the central subject.

The central government is not just a level in the bureaucracy; it is the central locus of moral authority. Beijing is not only a place; going to Beijing means facing *guojia* directly. In Zhao Liang's prize-winning documentary *Shangfang* (Petition), a veteran petitioner explained why she had not given up: "I didn't know there were so many petitioners when I was home. Once I came to *guojia*, I saw so many petitioners. Now I know there are so many petitioners."

What she meant by *guojia* was what we call Beijing or, more precisely, the area where the petitioners lived in makeshift shelters on the southern edge of the city (the so-called petitioners' village).³ By calling this place *guojia*, she was not trivializing the state. She was giving a physical location, a concrete contour, to the state in order to interact with it meaningfully. In counterdistinction to the Foucauldian notion that power manifests itself as widespread webs and penetrating capillaries, the ordinary people in China construct a notion of *guojia* as the central source and possessor of power. By doing so, they are not reifying or mystifying power. On the contrary, this is a strategy of making sense of, and, more importantly, countering, power.

The ministry officers were sympathetic and supportive. They confirmed that some activities of the FS government were indeed irregular: for instance, the Bureau of International Economy did not have the authority to issue the special permit to Green Metro. The victims were particularly pleased that the central state made it clear that the local government failed to follow existing rules, which they took as proof that the problem could and should be resolved *within* the administrative system, without any need to resort to legal channels. The victims raised questions and requests in a moralistic, right-or-wrong manner, and the central state similarly gave black-and-white answers. The victims felt that the central state was on the same page as they were.

Returning from these heartening visits to Beijing, the victims urged the government departments at the provincial and municipal levels for action again. The State Council Office for Letters and Visits also forwarded the victims' letter to the Liaoning provincial government, with a note that "relevant departments must take swift action," which was further handed down to the FS municipal government. The FS municipal government responded in writing that they would work out a "systematic method leading to a comprehensive solution."⁴ In these sorts of interactions, the central state secures its moral authority by blaming the local government. The local government, for its part, readily subjugates itself to the central authority when it is blamed, not only because the central state remains powerful (for instance, controlling personnel appointments), but also because the local government, whose own credibility is seriously undermined in daily interactions, derives its authority from being part of *guojia*. Local instrumentalization relies on the moralization of the center.

Eventually, the deputy mayor in charge of economic affairs assembled a meeting attended by members of the Office for Letters and Visits, Bureau of International Economy, Bureau of Public Security, and Bureau of Civil Affairs.

The mayor first of all expressed deep sympathy for the victims and mentioned that the top leaders of the municipality had asked about the case at least nine times. He then announced:

> The government had a meeting with nine departments. Each department expressed their attitude and determination (*biaotai, biaojuexin*) [to seek a solution]! But . . . it was decided that a lawsuit should be the main method for the next step. We do not rule out using multiple methods and approaches simultaneously. Finance-wise, we should focus on compensation; administratively, we can pursue with the administrative court; legally, we can pursue the case as a criminal offense associated with civic dispute (*minshi dai xinshi*). We should trust law, trust fact.

When Lin Feng uttered that they trusted the government more—he still preferred administrative interventions, the mayor stopped him immediately: "Government should also obey laws!" The victims were once again urged to sue Wang Luli for the criminal offense of deception and, if rejected, for contract violation (this was what the deputy mayor meant by "criminal offense associated with civic dispute"); they were also encouraged to sue the economic development zone at the administrative court.

By this point, the victims' campaign was beginning to lose momentum. As a last resort, they sued Wang Luli for breach of contract (they still avoided the criminal court because it would not help with compensation). At the same time, they sued the Bureau of International Economy, rather than the development zone, for negligence. This is because the bureau was higher up in the government hierarchy and was financially more resourceful. The court found Wang guilty but spared the bureau. As Wang proved himself to be incapable of compensating the victims, he was ordered to issue a Recognition of Debt, which obliged him to return the money whenever he was able to.

The victims were not particularly upset by the result. This was surprising to me, given that the victims had spent more than RMB 40,000 (USD 6,000) to hire their own lawyer for the case after turning down the free legal service from the government. Ma had no regrets:

> We feel the money was well spent! Our lawyer is very professional. And she really spoke for us! In the court she said everything clearly: one, two, three. She really said what we want to say from our heart! Wang Luli's lawyer couldn't answer her questions. He was just repeating the same sentences. . . . This lawsuit was the lodging of our spirit (*xinling de jituo*).

When there is a lodging of sprit, there is hope. Ma Na and I had the last meeting in a small KFC restaurant in her neighborhood. She was calm: "At the end of the day, that's [because] the system of law in our country is still not well developed. I just hope things will be better for the next generation." Ma had just had a baby daughter eight months earlier. I was particularly concerned about Lin Feng. When I visited him during my last trip in 2008, he was still deeply worried about the future—his whole family had moved to his parents' home in order to save on utility bills. But Lin appeared calm as well, and he drew an almost identical conclusion as Ma Na: "There is little you can do when the rule of law is so weak." Surprisingly, he expressed an even stronger sense of hope:

> This is after all a new problem that emerged in the course of social development.[5] Our *guojia* will certainly become better. Don't you think so? Just look what big strides *guojia* has made over the last years. Premier Wen first got rid of the rural levies, and now is providing free education. [Improvement in] medical care will also follow.

Hopeful Engagement and Political Subjectivity

A regime's capacity for generating a sense of hope is probably one of the most critical components of its actual existing legitimacy. It was also the sense of hope that enabled the victims to have come so far in pursuing their cause. As Lin Feng's hope demonstrates, people genuinely desire to establish a positive relation with the state. The people are not merely appropriating the words of the state for immediate benefits when they repeat the official lines. They are engaging in dialogue with the state. This desire for engagement and the sense of hopefulness differ from the structural chasm in the traditional Chinese political imaginary that the center embodied the heavenly mandate (*tianming*) while the local officials were corrupt. Although traditional idioms such as the Blue Sky Bao play a role in the contemporary political expression, and petition has a long tradition, the experiential basis of the current political morality is very different. The world where "heaven is high, and the emperor is far away," which had helped sustain an idealist image of the center and reconcile it with corrupt local officials, no longer exists; the three decades of state socialism after 1949 fostered intimate relations, both sweet and bitter, with both attachment and fear, between the people and the state. The ordinary people feel that they not only "owe" the state, but that they also "own" it. They feel they owe *guojia* as much as *guojia* owes them. Thus, the moralization is not a continuation of the political cosmology that prevailed in imperial China, but is rather part of what I call the "folk theory of the state" that is specific to contemporary China (Biao Xiang 2010). By "folk theory of the state," I mean people's relatively systematic reasoning as well as normative expectations about the state's role. The crossing between the instrumentalization of local life and the moralization of *guojia* is not tactical maneuvering between public transcript and hidden transcript as a weapon of the weak (Scott 1990). It is a weapon of the confident, the demanding, and the loud.

This political morality is also different from the mainstream ideology in liberal democracies. For instance, the Chinese people commonly present themselves as subjects (*laobaixing*) with entitlements instead of as citizens with rights, and they judge the state according to its substantive deliverables rather than the formal procedures to which it adheres. This may appear transitional or even pathological if we take liberal democracy as the standard, but this can also be understood as the people's strategy of retaining their political subjectivity and refusal of becoming liberal legal-economic subjects. Without such refusal it would be impossible to translate commercial disputes into incidents over which people can make political claims and moral demands.

As one of the most important legacies of socialism, such political subjectivity is deeply rooted though not necessarily well articulated. This is perhaps one of the most important political resources that we have for now. People see themselves as agents who are capable of taking political actions, while legal procedures are relativized, and market principles are subject to negotiation, even though legalism and developmentalism appear to be unquestionable hegemony. Such a political subjectivity may well be a central driving force for China's sociopolitical transformation. If the leadership is responsive enough to the people's political subjectivity, China could achieve fundamental yet relatively peaceful transformation toward a more democratic and accountable regime. But if the political subjectivity is suppressed or antagonized, for instance when the relationship between the people and the central state becomes instrumentalized or there is no longer a sense of hope, China may embrace revolution. This is a time when despairs and hopes entangle, and the burning desire for change and chilling fears about disorder intertwine. Most people feel that deep changes may be imminent, yet almost no one can envisage a road map for those changes. History may proceed in different directions, and we must therefore be able to *envision* possible futures. In order to do so, it is an urgent task to articulate, reflect on, and engage with the people's political subjectivity.

Notes

1 According to a survey conducted by a Taiwan-based project conducted in Mainland China in 2008, nearly 80 percent of those surveyed strongly agreed or somewhat agreed with the statements "Whatever its faults may be, our political system is still the best for the country's current condition," and "Even if we don't agree with the government's specific policy, once the decision is made we should still support the government" (Chu 2011: 37–38).

2 Clothes and medicine are the two most important items that would-be migrants prepare before working overseas. Migrants typically bring a large amount of medicine because they often fall ill due to harsh working conditions yet they cannot visit doctors on account of their tight working schedules.

3 The "petition village" was a spontaneously formed concentration of petitioners in south Beijing close to the Supreme Court and the petition offices of the Office of the State Council and the Office of the Central Committee of the CCP. Petitioners lived in shelters or simply under bridges—some did so for many years—until the village was demolished in 2008. This petition village was also where most parts of the documentary film *Shangfang* (Petition) was shot.

4 Intragovernmental communication like this is supposed to be highly confidential, but the victims somehow obtained a copy. Ma Na was reluctant to reveal how exactly that happened, but it is clear that the boundary between the local state and the society is constantly transgressed.

5 This is a familiar official line, the typical expression of which is "[this] is a problem that emerged during the reform and should be dealt with through deeper reforms."

7

QUEER REFLECTIONS AND RECURSION IN HOMOEROTIC BILDUNGSROMAN

Rachel Leng

When a tabloid reporter writes an exposé on the tavern the Cozy Nest near the end of Pai Hsien-yung's novel *Niezi* (Crystal boys), the establishment's delicate social dynamic as a gay meeting ground is disrupted. The Cozy Nest soon becomes much less cozy after it is flooded with curious visitors who regard Taipei's young male sex workers as objects of fascination. The reporter reveals more details about the tavern in subsequent articles, whereupon the establishment—which had served as a quasi-secret meeting place for these young men and their male clients even while operating in plain sight as a public business—is ultimately forced to close because its existence has been revealed to the wider public.

Published in Chinese in 1983, *Crystal Boys* is often cited as the first modern Chinese-language novel to focus on issues of same-sex desire, and there is a sense in which the work itself reflects a twist on the logic outlined in the novel's own Cozy Nest sequence. That is to say, like his fictional reporter, Pai Hsien-yung based his novel on Taipei's contemporary gay community, and part of the work's subsequent popularity was a result of the way it helped make this community visible to a broader readership. Unlike the scenario that unfolds in the novel's Cozy Nest sequence, however, the work's own success in drawing public attention to Taipei's gay community has not resulted in the community's dissolution, but rather has contributed to its increased visibility and long-term vitality.

In particular, the global influence of gay activist groups in the wake of the HIV/AIDS pandemic that began in the 1980s, combined with the process of cultural liberalization in Taiwan following the lifting of Martial Law regime in 1987, has contributed to the visibility and vitality of the nation's queer community. For instance, one index of this surge of attention to queer issues may be found in the literary realm, and Pai's *Crystal Boys* was followed in the 1990s by works such as Ling Yan's *Shisheng huamei* (The silent thrush), about an all-female Taiwanese opera troupe; Qiu Miaojin's novels and short stories, which explore lesbian concerns; and Chu T'ien-wen's 1994 novel *Huangren shouji* (Notes of a desolate man), which is narrated in the voice of a gay Taiwanese man mourning the recent AIDS-related death of a close friend. Written by both queer and straight authors, these literary explorations of queer identity and same-sex practice in Taiwan have proceeded hand-in-hand with shifting attitudes about homoeroticism, including those by self-identified members of Taiwan's LGBT community themselves.

In Mainland China, meanwhile, there was a similar surge in homoerotically themed literature during this same period, though in response to a somewhat different configuration of cultural and sociopolitical factors, including the process of liberalization resulting from the 1980s Reform and Opening Up Campaign. Representative works include Lin Bai's 1994 novel *Yigeren de zhanzheng* (A war of one's own) and Chen Ran's 1996 novel *Siren shenghuo* (A private life), both of which prominently feature lesbianism, as well as Cui Zi'en's 1998 novel *Choujiao dongchang* (Enter the clown) and other works, which feature male homoeroticism, incest, and transsexuality. While all of these works were published through official channels (though some of Cui Zi'en's works were not approved for publication in China), the explosive growth of the Chinese Internet has also had a transformative influence on this burgeoning body of literature on same-sex desire, in that it offered a sphere where works could be published without having to go through the official censorship apparatus. The result has been the emergence of an entire genre of literary works disseminated through Chinese-language websites, many of which have millions of subscribers.[1]

By 2005, there were more than five hundred Chinese websites publishing homoerotic-themed content (Cui 2008). As authors take to the Internet to discuss formerly taboo topics, they help destabilize preconceived notions of gender and sexuality, while generating new queer ones. Their texts articulate private desires within a public space, highlighting central concerns regarding the relationship between social oppression and the expression of sexual identity in contemporary China (Berry, Martin, and Yue 2003). These works dramatize a process by which gays in China navigate their individual identity in a nation influenced by a growing attitude of sexual openness that nevertheless continues to be inflected by the institutional and ideological vestiges of a more restrictive past.

The ability of online homoerotic fiction to explore private—and often socially proscribed—topics within a paradigmatically public space mirrors the status of many Chinese queers (or *tongzhi*, as they are commonly called) themselves. In China, the *tongzhi* community is nearly invisible to the general public, and *tongzhi* groups cannot publicize themselves as organizations associated with homosexuality; instead they must register as HIV/AIDS institutes or social groups such as book clubs (van der Werff 2010). Although many Westerners remark on the growing number of gay bars and clubs in Beijing and other major cities, Chinese *tongzhi* themselves lament that these seemingly liberating semipublic areas are ultimately just "enlarged closets" to which they remain confined. Most of them resent the misconceptions caused by a lack of information and are discouraged by the fact that, despite contemporary China's rapid transformation, they nevertheless remain comparatively marginalized (Fang 1995). Rather than remaining in their hometowns, accordingly, many *tongzhi* choose to migrate to megacities like Beijing and Shanghai, to seek communities that are potentially more open and tolerant (Ho 2010). In this way, *tongzhi* hope to leave behind conservative neighborhoods and instead pursue economic and romantic opportunities in the metropolis, where they may also enjoy greater anonymity (Cui 2008).

In this chapter, I use the 1999 Internet novella *Huizi* to examine a related set of issues of intimacy, self-expression, community formation, and sexual identity. Published under the pseudonym Xiaohe,[2] *Huizi* depicts the relationship between two boys as they mature and become aware of their same-sex desires, and the result is a work that brings together a set of parallel "coming-of-age" and "coming-out" narrative conventions. In particular, the novel focuses on the relationship between the narrator, Xiaoyang, and his best friend, Huizi. While Xiaoyang remains on good terms with his family and his community, Huizi becomes increasingly ostracized within the community as he begins to acknowledge his homoerotic yearnings. Within the novel he comes to symbolize not only a certain queer positionality, but also a persistent homophobia that continues to be found in contemporary Chinese society.

In *Bodies That Matter*, Judith Butler argues that certain bodies are abjected by a process wherein the "materialization of the norm in bodily formation produce[s] a domain of abjected bodies" that "[fail] to qualify as fully human" and are not recognized as valued subjects in contemporary culture. According to Butler, these bodies represent a "constitutive outside" of the social order in that they are "delegitimated bodies that fail to count as 'bodies'" (1993: 16, 15). In *Huizi*, Xiaoyang's relationship with his friend is one wherein Huizi reflects both Xiaoyang's own burgeoning homoerotic desires, as well as the prejudicial attitude with which that same homoerotic desire is frequently viewed in contemporary China. As a result, Huizi functions as a figure of abjection—reflecting both the abject position occupied by queers within contemporary Chinese society, as well as the degree to which his friend Xiaoyang has internalized those same prejudices even as he struggles to come to terms with his own sexuality.

In this way, *Huizi* offers a framework for reexamining modern China's rapidly changing notions of social norms and sexual identity, specifically the ways in which an increasingly liberalized political order draws on vestiges of an earlier, less tolerant regime to generate a set of conditions that constrains but simultaneously encourages the emergence of new subject positions. That is to say, *Huizi* illustrates how, both at an individual and at a societal level, one finds a dialectical relationship between growing acceptance of homoeroticism, on one hand, and the vestigial legacies of earlier, less tolerant regimes, on the other. I argue that these legacies of prejudicial attitudes are manifested in the form of figures of abjection, which in turn come to play a productive role in the constitution of new queer subject positions.

Huizi or Hooligan? The Imbrications of Language and Social Identity Formation

Huizi opens with Xiaoyang explaining that "*Huizi* was my nickname for Li Zhanghui, while 'hooligan' (*liumang*) is the status to which everybody ultimately condemned him. Huizi and I grew up together in the same neighborhood."[3] *Huizi* could be translated as "splendid son" or "radiant boy," while *liumang* is commonly used in China's recent penal code to refer to those who engage in same-sex practices. The novel's juxtaposition of these two terms, accordingly, reflects a pair of markedly different attitudes toward same-sex desire in contemporary China, wherein growing tolerance continues to be haunted by a set of entrenched societal prejudices.

Under Mao, China attempted to eliminate all nonprocreative sex in order to establish the nuclear family as the bedrock of socialist renewal (Sieber 2001). The 1957 Official Penal Code included what came to be known as the "hooligan" law (*liumangzui*), which listed sodomy as a criminal act, and in 1984 homosexuality was classified as a pathology in the first Chinese *Classification and Diagnostic Criteria of Mental Diseases* (Mountford 2010). As a result, anyone suspected of same-sex behavior could be arrested and condemned to a labor reform camp, an electric-therapy clinic, or even execution (Wan 2001). Currently, though, China formally neither prohibits nor protects same-sex practice. The abolishment of the "hooligan" law in 1997 decriminalized homosexuality, and in 2001 homosexuality was removed from the official list of mental disorders (Mountford 2010). The Chinese state officially acknowledged the presence of homosexuals for the first time in 2004, when a government survey placed the nation's homosexual population at about fifteen million, though more recent estimates have ranged from fifty million to one hundred million (Gong 2009).

Some of the tensions inherent in this growing visibility and activism on the part of China's lesbian and gay community are reflected in the term that many Chinese gays and lesbians have adopted to describe themselves: *tongzhi*. The term means "same will" or "of the same intent" and was first popularized in a modern context by the Nationalist leader Sun Yat-sen, who used it to refer to those who shared his revolutionary aspirations. The word was subsequently adopted by the communists to express their shared goal in building a socialist China, and even today it continues to carry strong Maoist connotations (Scotton and Zhu 1983).[4] By referring to themselves as *tongzhi*, Chinese gays effectively queered one of the most revered terms from the same contemporary political regime under which they had been oppressed. The result figuratively reimagined China's revolutionary history and helped break down the ideological barriers between the nation's socialist past and its putatively more liberal present (Wong and Zhang 2000).

In contemporary China, meanwhile, the state continues to view "sexual openness" (*xing kaifang*) as a threat to socialist morality and sociopolitical stability, and consequently several legal statutes continue to target homosexuals (Sigley 2006: 71). The Social Order Statute and laws on harmful sexual acts are still used to detain homosexuals, criminalizing behaviors that are "deleterious to respectable customs" (*fanghai shanliang fengsu*) or "deleterious to moral decency" (*fanghai fenghua*), and dictating that "diseased" (*bingtai*) or "abnormal" (*biantai*) citizens be detained (Mountford 2010). In general, however, the government's preferred tactic has been to act as though homosexuality does not exist—an approach that is characterized as "not encouraging, not discouraging, and not promoting" (*buzhichi, bufandui, butichang*). In practice, though, this policy is not neutral and instead creates an environment where homosexuals are relegated to a legal twilight zone (Gong 2009).

These conflicted attitudes toward queer desire in contemporary China provide the backdrop against which the plot of *Huizi* unfolds. Xiaoyang and Huizi begin as childhood best friends, and Xiaoyang subsequently comes to view himself through his adolescent friendship with Huizi, who functions as both an object of desire as well as a site of projective identification. In this respect, Xiaoyang's relationship to Huizi is reminiscent of the way in which an infant, as Jacques Lacan argues in his influential mirror-stage essay ([1949] 2002), comes to acquire a sense of self through a process of "jubilantly" identifying with an image in a mirror, and while Lacan's essay focuses on a specific moment in infancy, this model of projective identification may also be applied to a broader range of phenomena wherein individuals model their self-identity on an external sociocultural image. But what happens when this external image is one that is explicitly devalorized? In *Huizi*, as Xiaoyang becomes aware of how his sexual orientation is at odds with Chinese heteronormative ideals, his desire to identify with Huizi—and, by extension, with a *tongzhi* identity—becomes increasingly difficult, ultimately leading not to jubilation but to bitter dejection.

This tension between processes of identification and abjection is exacerbated after Huizi is arrested and imprisoned. After school one day, Huizi witnesses a fight in which another boy is killed, and although Huizi was not directly involved he was nevertheless convicted on charges of "hooliganism." Xiaoyang describes the dramatic changes that occurred overnight:

> This event shocked everyone.... Nobody ever imagined that the well-mannered and promising boy from the Li family would become a "little hooligan." In the space of a single evening, Huizi's father never smiled again, and instead came to have as much vitality as a block of frost. My own father claimed he felt sorry for Huizi, but when he spoke I could see ridges of delight along his forehead. (2)

Xiaoyang's observation that his father took delight in Huizi's misfortune underscores the emotional rift that exists between Xiaoyang and his parents—suggesting that "hooliganism" may contribute to feelings of estrangement and alienation that eventually undermine the family itself. While Xiaoyang is troubled by the injustice of Huizi's imprisonment, his parents and the other adults in the neighborhood simply accept Huizi's charge of hooliganism without question. The irony is that even though the "hooligan" law was ostensibly enacted in an attempt to uphold Confucian values of the nuclear family, in Huizi's case this charge for which he was imprisoned contributes to his growing alienation from his own family.

Xiaoyang is raised in an environment that emphasizes filial piety, where boys were expected to grow up, marry, and have sons to continue the patriline. Huizi, by contrast, finds himself increasingly marginalized, and while his initial imprisonment was not directly related to his sexuality, the specific charge of "hooliganism" directly echoes the language used during that period to prosecute homoerotic activity. Although Huizi is released from prison early for good behavior, his criminal record nevertheless continues to haunt him. He is barred from returning to school or pursuing a respectable career, and his own parents brand him a "rotten hooligan" (*chouliumang*) and someone who "plays with hooliganism" (*shua liumang*). This backdrop of legal and social discrimination reflects the fluctuating set of approaches to the label of hooliganism and disciplinary regimes in contemporary China.

At one point, Huizi's mother lashes out at him for chatting with a girl in the courtyard, calling the two of them "rotten hooligans" and telling them to get lost (2). After Xiaoyang witnesses this episode around the same time that he begins to identify with Huizi, he reacts defensively by withdrawing from interpersonal relations. Huizi's abjected "hooligan" body thereby becomes a figure of alterity against which Xiaoyang positions himself. In the process, Xiaoyang attempts to claim an identity that is as different from Huizi's as possible, distinguishing himself as "a model student who doesn't fight, studies hard, and doesn't have romantic relationships" (2). However, when Xiaoyang finds himself unable to establish an emotional connection with others, this absence of inner trustworthiness is projected onto an outside world that is perceived as harsh, desolate, and uncaring.

Xiaoyang's urge to attach himself to Huizi, accordingly, can be read through the lens of Lacan's description of the infant's projective identification with its mirror image, or *imago*. However, Xiaoyang's subjectivity alternates between jubilant recognition with and a paranoid dissociation from this external *imago*, leading to a fragmented state of alienation. Xiaoyang's predicament mirrors the broader sense of displacement that *tongzhi* individuals experience in contemporary China. On one hand, Chinese *tongzhi* face a restrictive society where deviating from an imagined social norm is characterized as "hooliganism" and individuals are not provided a space for public self-expression, but on the other hand the very process of labeling individuals as "hooligans" forges a space at the margins of society from which new counterhegemonic subject positions may emerge.

Butler argues that the reconsideration of an abjected body is a crucial political move that permits a renegotiation of who or what counts as a body that "matters" (1993: 2). She refers to this as the "politicization of abjection" where "the public assertion of 'queerness' enacts performativity as citationality for the purposes of resignifying the abjection of homosexuality into defiance and legitimacy" (1993: 21). The application of the hooligan label to figures such as Huizi may be viewed as a process of "reworking . . . abjection into political agency," wherein *tongzhi* characters confront the nuances of personal experiences in contemporary Chinese society's neoliberal social norms through (mis)identification with an abject body (Butler 1993: 21).

One good example of this reworking of abjection can be found in a passage where Huizi teases Xiaoyang for expressing concern about Huizi's experiences while in prison. Huizi jokes that Xiaoyang is acting like his "wife" (*xifu*), to which Xiaoyang replies, "What's wrong with being your wife? If you are willing to have me, I'm willing to be one!" (3). Huizi then proceeds to exhort Xiaoyang "not to become bad" (*bie xue buhao*), whereupon Xiaoyang realizes with a shock that Huizi was interpreting Xiaoyang's offer to be a "wife" in sexual terms:

> That conversation is one I will never forget for as long as I live—it was like an alarm, brutally forcing me awake with the recognition that I am, in fact, "becoming bad"! For the first time, I was shocked, confused, distressed, and even terrified by my feelings for Huizi. When I later reflected on that dialogue, I realized that Huizi and I had very different understandings of what it means to be a wife. I was thinking of an emotional bond, but Huizi was evidently thinking of sex. (3)

By presenting alternate understandings of what it means to be a "wife," this passage deploys a tension between differing conceptions of gender and sexuality, and in the process it reveals a set of ideological rifts produced as conservative social norms are transformed and reinvented.

More generally, this passage illustrates the novel's concerns not only with how *tongzhi* are regarded by others, but also with how they perceive themselves. Throughout the work, Xiaoyang feels increasingly apprehensive about his own sexual inclinations, even as he is captivated by Huizi's experiences in prison and how they have shaped his body. By desiring Huizi because of his abject "hooligan" body, Xiaoyang seeks not only to understand his homoerotic feelings for Huizi, but also to define his own sexual identity in an environment where the spectral residue of China's history of criminalizing and pathologizing homosexuality still looms large.

Xiaoyang's desire for Huizi is contrasted with the emotional distance that increasingly separates the two young men. Xiaoyang feels that Huizi has no intention of returning his affection—especially after Huizi starts dating another boy named Xiaowei. Devastated, Xiaoyang decides to attend a university in Shanghai to get away from Huizi, and in one pivotal scene he attempts to purge himself of his same-sex desire:

> I started to caress my body, and imagined that these were Huizi's hands I felt. . . . I whispered his name and felt a weight on my body, as if he were lying on top of me. I gazed at him in joy and with my hand I stroked his handsome face. . . . But there was no Huizi; his hands, his lips, his caresses were never there. My hands were only full of a viscous fluid. I stood up to clean off my own semen, and I assiduously used toilet paper to wipe my hands but could not get them clean, as bits of paper kept sticking to my hands. I wiped harder! Harder!
>
> . . . I walked out the house into the yard, turned on the water pump, bent over, and put my head under the gushing water. . . . The ice-cold clear water slowly washed away the dry summer heat. The more water gushed out, the colder it became, until it became so cold that it hurt. . . . But I didn't want to stop; I needed the pain. I wanted to completely wash away the body fluid from my hands, the tears from my face, and the filthy thoughts from my mind.
>
> I started my life as a university student after rinsing my filthy body with that icy water. I was so naïve to think that the tormenting stream of water could wash away my love for Huizi and my desire for men. Who knew the water of Beijing would turn out to be so unreliable? (4)

In this scene, Xiaoyang shifts abruptly from a fantasy of erotic fulfillment to an alienated rejection of those same desires. His desperate efforts to cleanse himself of the material evidence of those desires, however, come to function as a new object of desire in their own right.

Xiaoyang's semen, in this passage, may be viewed as a paradigmatic example of what Julia Kristeva calls the abject, or "horrors [from] within" that "show[s] up in order to reassure a subject that is lacking its 'own and clean self.' The abjection of . . . flows from within suddenly become[s] the sole 'object' of sexual desire—a true 'ab-ject'" (Kristeva 1982: 54). Thus, the ejaculate symbolizes the collapse of an imagined border between Xiaoyang's homoerotic desires and external heteronormative reality. In this sense, Xiaoyang's dismay upon realizing that "there was no Huizi" and that his "hands were only full of a viscous liquid" illustrates his sense of sexual abjection (4). The ejaculate deriving from Xiaoyang's fantasy about Huizi materializes the abject homoerotic desire that Xiaoyang is simultaneously trying to exorcize—but in the process it ironically ends up reinforcing that same desire as a fundamental aspect of his subjectivity (4).

The ejaculate provides a provocative closing image to this chapter of the novel, crystallizing not just the psychic aspects of Xiaoyang's homoerotic desire and self-abjection but also its constitutive (homo)sexual foreclosure. Xiaoyang's confrontation with his own semen signifies the nonprocreative *tongzhi* body as a site where hope for rebirth and reproduction is short-circuited. The existence of the semen suggests that Xiaoyang's attempt to purge his homoerotic desire through the act of indulging it underscores the fact that he appears unable to recognize the *tongzhi* identity that he has already unconsciously acknowledged. Through Xiaoyang's simultaneous affirmation and disavowal of his same-sex desires, it is possible to understand how he can simultaneously exist yet not exist as a (self-denying) gay man.

Xiaoyang decides to leave Beijing to distance himself from Huizi and his homoerotic desires, and in Shanghai he begins to actively explore his sexuality. He initially makes a point of dating women, but these relationships invariably fail, and instead of enabling Xiaoyang to lead a "seemingly 'normal' life," these heterosexual relationships only exacerbate his sense of self-abjection (8). He repeatedly expresses concern that his girlfriend might discover his "vulgar inner self," and he worries that "if [he] were to continue dating [them], the secret in [his] heart would no doubt be exposed" (8). The paradox here is that the very thing that promises Xiaoyang a "normal" heterosexual life underscores his lack of desire for women, signifying the very opposite of male heteronormalcy in patriarchal Chinese society.

Between his unfulfilling relationships with women, Xiaoyang meets up with other men in public toilets for anonymous sex, and the result of these sexual exchanges is a complex space marked by the movements of people who wander the city in search of others who are themselves in a state of continual circulation. After graduating from college, Xiaoyang conforms to heteronormative expectations and marries a female classmate. He settles down in a small town for several years but returns to Beijing upon receiving a letter from his parents reporting that Huizi had been sent back to re-education labor camp for engaging in "hooligan behavior." Upon visiting the camp, Xiaoyang is confronted with the sight of an emaciated Huizi:

> I felt an uncontrollable agitation; facing the person I loved, facing his languished silhouette, I could not simply feign indifference. From his emaciated face, cold gaze, and tightly pursed lips, I could see Huizi's pain, and that pain made me tremble. Not wanting him to see the tears in my eyes, I immediately lowered my head.... After I stabilized my emotions and looked up, Huizi was actually smiling, though his smile still revealed bitterness. (9)

The spectacle of Huizi's wasted body causes Xiaoyang to feel an acute sense of grief, even revulsion. Xiaoyang sees his *tongzhi* identity reflected through the figure of Huizi, but this relationship is shattered when Xiaoyang is confronted with Huizi's abject body, forcing him to recognize their estrangement.

Upon leaving Huizi, Xiaoyang begins to vomit—a reaction that may be seen as another example of Kristevan abjection, which she describes as a process whereby

> I expel *myself*, I spit *myself out*, I abject *myself* within the same motion through which "I" claim to establish *myself*. That detail... turns me inside out, guts sprawling; it is thus that *they* see that "I" am in the process of becoming an other at the expense of my own death. During that course in which "I" become, I give birth to myself amid the violence of sobs, of vomit. (Kristeva 1982: 3; emphasis in original)

Huizi's character serves as a figure of desire and abjection, and Xiaoyang's response to his encounter with him displays the traumatic experience of being confronted with a materiality that signifies his own homosocial death and "show[s] [him] what [he] permanently thrust aside in order to live" (Kristeva 1982: 3). Nonetheless, Xiaoyang's final act in the story is to ensure the man he loves is taken care of in prison, as he persuades Huizi's sister to accept her brother's homosexuality and reunites Huizi with his earlier partner, Xiaowei. In this way, Xiaoyang accepts his role as a bereaved lover and in the process appears to both repress his homoerotic desires and reaffirm his own heterosexual marriage.

One day several months after Xiaoyang returns home, however, he finds himself sitting alone and playing with the gold ring Huizi gave him on his wedding day. When looking at the character for "fortune" (福) engraved on the ring, he ponders the blessings Huizi had given him. Before long, Xiaoyang's wife comes up behind him and disrupts his reverie by dragging him from his chair for sex. The novel ends with a description of Xiaoyang "unable to resist looking back at the ring . . . with 'fortune' shining under the table light," remarking that "it was as if its golden yellow blessings were radiating throughout the entire house" (10).

In looking back at the ring, Xiaoyang reflects on the suffering Huizi has endured, while at the same time envisioning the possibility of a more progressive and tolerant future. The ring, therefore, figuratively sutures together Xiaoyang's past, present, and future, thereby helping him negotiate his relationship to his wife and others around him. It symbolizes both Xiaoyang's nostalgia for same-sex intimacy as well as his current heterosexual marriage, while at the same time it marks a final return to the work's concerns with the dialectical relationship between secrecy and disclosure. By forging a link between inner and outer experiences, the ring connects Xiaoyang's private homoerotic fantasies with his public heteronormative reality. Xiaoyang is thus psychically and physically detached from Huizi, even as he remains symbolically connected to him. A widely recognized symbol of love and romantic commitment, the ring also represents how Xiaoyang's lack of fulfillment in his heterosexual marriage ensures the persistence of his homosocial desire.

At the same time, Xiaoyang's wedding band may also be treated as a symbol of the position of online homoerotic literature, like *Huizi* itself, in contemporary China. Like the wedding band, this literature has overlapping public and private dimensions, in that it focuses on paradigmatically private desires and intimacies yet does so in a medium explicitly designed for public circulation. In this respect, these stories are symptomatic of the liminal position that queer subjects occupy in urban China, under the state's constantly shifting disciplinary and surveillance regimes. The comparative accessibility of these Internet texts serves as a tacit reminder of the degree of oversight the government continues to maintain over more formal channels of publication and communication, together with the way in which those latter restrictions may paradoxically offer a stimulus for the sorts of informal domains of cultural production found on the Internet and elsewhere.

Virtual Communities

Electronic technology enables an innovative way of writing and invites a new understanding of authorship (Zhang 2011), wherein the production of Internet *tongzhi* literature becomes a community-building activity emblematic of

Chinese online struggles for recognition and against discrimination. The participatory landscape of virtual space is exemplified by features such as relay writing, a process whereby *tongzhi* authors receive reader feedback during the writing process (C.S.M. 2013).[5] This interactive process is evident in the serialized posting of *Huizi*, where the author left comments at the end of each chapter, responding to reader feedback.[6] Some fans went further and composed entire epilogues or sequels to the novel, many of which concluded with Xiaoyang and Huizi ending up as a couple. In this way, the social and institutionalized homophobia that is foregrounded within the original novel helps generate its very antithesis, in the form of a same-sex romance that, while fictional, arguably has practical implications for the virtual communities of authors and readers within which it circulates. Online *tongzhi* literature enables participants to be both authors and readers, inspiring the sharing of memories and sentiments underrepresented in China's (post)totalitarian public sphere and producing a virtual community wherein *everybody* may be viewed as *tongzhi*.

In this way, *tongzhi* literature represents a possible renegotiation of paradigms of gender and sexuality, particularly insofar as they are shaped by sociopolitical forces. As such, the online *tongzhi* platform functions as a rhizomatic public space in which private narratives and common experiences may be copied, compared, and circulated. Thus, *tongzhi* stories such as *Huizi* offer poignant accounts of homoerotic romance and desire, but they also reflect the continued denigration of *tongzhi* existence in modern China. This narrative trajectory exposes the lingering effects of abjection highlighting the gap between homosocial aspirations and heteronormative realities amid shuffling social structures in contemporary China. *Tongzhi* coming-of-age narratives like *Huizi* reflect a set of conflicting attitudes toward homoeroticism in contemporary China, even as they simultaneously permit readers to transcend abjection of the nonheteronormative body consigned to the margins of society, edging toward expanding the future horizon of coming out to *tongzhi* potentiality.

Notes

1 For example, as of November 2012, BoySky (*yangguang didai*) had more than seven million registered users with an average of 587 page visits every day. As a prominent *tongzhi* website, BoySky currently holds more than twenty thousand stories, many of which are updated regularly. Approximately five to ten new chapter additions are posted per day ("BoySky Homepage," www.boysky.com).

2 Most Chinese Comrade stories from this period were published either anonymously or pseudonymously.
3 *Huizi*, chapter 1. Published on the Internet, *Huizi* is not paginated, though it is divided into numbered chapters. Subsequent citations from the novel are followed by a reference, in parentheses, to the chapter in which the passage appears. All translations are the author's own.
4 In 1989, *tongzhi* was used in the Chinese title of the inaugural Hong Kong Gay and Lesbian Film Festival (*xianggang diyi jie tongzhi dianying jie*), and it was subsequently adopted by gay and lesbian organizations in Hong Kong and also exported to Taiwan, Mainland China, and diasporic Chinese communities (Chou 2000). *Tongzhi* quickly became the most popular term for Chinese homosexuals, especially gay men. In contrast, the medical term *tongxinglian* (homosexual) is rarely used within the queer community, because it carries connotations of mental disease (Kong 2011: 14).
5 The participatory nature of online Chinese fiction has been noted in a number of studies. For examples, see Lugg (2011) and McDougall (2003).
6 See a version of the serialized posting here: http://tieba.baidu.com/p/10902400?pn=1.

PART III

MIGRATION AND

SHIFTING IDENTITIES

8

TEMPORAL-SPATIAL MIGRATION

Workers in Transnational Supply-Chain Factories

Lisa Rofel

How do people make a transition from one world to another? In some places and times, transformation is so imperceptible it barely feels noticeable, but in others, such as in contemporary China, virtually everyone, except the very young, has lived through life-altering transformations affecting everything from their experiences of their own bodies to their dreams of what a future might look like. I refer, of course, to what is called the transition from socialism to capitalism. The socialist era had been fraught with contradictory tendencies, as periods of social and economic upheavals were punctuated by periods of economic growth. These latter periods of growth were based on a modernist program of development that mirrored capitalist ideologies, with the important exception of an emphasis on state-led redistribution of the fruits of production. The ideological commitments to egalitarianism were entangled with the production of new inequalities. Thus, the transition to a capitalist form of economic development contains some stark contrasts with the socialist era while also extending the latter's developmentalist tendencies.

One of the most discussed aspects of this transition is its attendant social inequalities. In public discourse and private conversation alike, people debate, assess, defend, and denounce these inequalities. And while inequalities of political power—and hence privilege—certainly existed during the socialist era, the tremendous widening of the differences in wealth and poverty—which in turn now define future possibilities and constraints—is a striking contrast with the previous era. To point to the immanent logics of capitalism or the overwhelming power of the state to understand this transition is insufficient. Rather, we need to highlight how people from heterogeneous life experiences embrace and produce new modes of being in historically contingent, sometimes surprising ways. The "protocols" discussed in the introduction to this volume highlight this historical contingency, what I have elsewhere called the experimental approach to neoliberalism (Rofel 2007).

I bring this question of historical transformation to bear on the lives of workers in contemporary China, specifically for those who work in factories geared toward the export of Italian fashion clothing. What does it mean for these workers, the overwhelming majority of whom are migrants from the countryside, to labor in these factories? Does this mark a significant transition from a rural life to an urban or semi-urban one? Do they experience it as a clear divide between socialism and capitalism? Or is it just another subaltern experience of marginal and marginalized living? What other transitions might it signify? To turn the question around, to what extent do factories located within the supply chain of Italian fashion have a specific mode of mediating migrant workers' experiences of inequality (in its broadest sense), their moral valuation of their labor, and their desires—all of which have been shaped in the aftermath of what is generally labeled the postsocialist era, which includes China's entry into the World Trade Organization and its subsequent intensive capitalist take-off? How might this experience contribute to producing alternative modes of being? Do workers dream of other futures? If so, which ones?

As stated in the introduction to this volume, the transformations in China since the mid-1980s have been monumental. While the state has gradually retreated from a centrally planned economy, it has continued to be involved intimately in the means and modes of profit seeking. It would be misleading, then, to characterize the market economy in China as in opposition to the state. We have witnessed the marketization of power, inequalities in distribution, rent-seeking behavior, increasingly polarized income levels, privatization of public resources, the abolition of security in employment, and the end of many social benefits formerly distributed by the state. The liberalization of foreign direct investment (FDI) has also become a driving force. China now has the greatest share of FDI in the world, surpassing the United States.

A recent special section of *South Atlantic Quarterly* (SAQ), edited by Ralph Litzinger (2013), describes the central role played by multinational corporations, with their "race to the bottom" production strategy, in the horrendous working conditions of today's migrant workers in China. Indeed, these transnational relations are a key part of the puzzle of how workers experience this historical transformation. The involvement of these corporations, however, is both revealed and obscured by the nature of the commodity chain, with its multiple intermediaries. Litzinger highlights the continuous blurring of truth and fiction in the way that multinationals such as Apple pride themselves on their corporate responsibility. Italian fashion companies certainly participate in transnational supply chains and the difficult working conditions faced by migrant workers in the textile factories. They also avoid direct responsibility by going through intermediaries. However, it is also important to be specific about which industrial production sectors one is describing. Litzinger and Pun and Chan in the *SAQ* special issue focus on relatively new electronics factories. The textile industry, in contrast, has a long history in China; it formed some of the initial basis of the socialist economy and provided labor and social stability for the new socialist government, especially in urban areas. It is a labor-intensive industry that also has a well-developed domestic sector. Thus, migrant workers in the textile industry can move about among different kinds of textile factories, some with working conditions that are better than others. The dyeing factories certainly deal with unhealthy chemicals, the spinning and garment factories less so. But all have back-breaking work schedules and insecure working conditions. Indeed, as I argue later, textile factories for export fall under the radar of activist organizing that highlights sweatshop conditions in China. In textile factories for export, the issue is less constant overwork or even horrible factory conditions than the risk and insecurity workers face in relation to the uncertain and ever-changing schedules associated with the export of fashion garments. Workers in this sector are essentially being taught to be entrepreneurial about their factory labor—to accept this risk and uncertainty as part of the very conditions of their labor.

Yan Hairong (2008) has argued that migrant labor is vital to China's flexible accumulation and growth, and in this chapter I contend that migrant workers' encounters with factory labor in Italian fashion production are informed by the dynamic intersection of three forces: namely, origin stories, affective engagements with temporality, and transnational encounters. I argue that these three forces, articulated together, lie at the heart of how people move from one world to another and shape how they interpret and grapple with social inequalities.

In anthropology, we have long paid attention to what the discipline calls origin stories. Origin stories, anthropologists have argued, are allegories that social groups tell themselves about how they have come to be in the situation they are in. Sometimes they also point toward unfinished matters and ambiguous resolutions to dilemmas. Although I do not have space in this chapter to elaborate on contrasting the origin stories of workers, elite entrepreneurs, and government officials involved in the textile and garment export business, it is clear that each group has a distinctive manner of invoking the past. Elite entrepreneurs reject the socialist past tout court, while they have developed a nostalgia for a revisionist imagination about the presocialist past, especially 1930s Shanghai. They invoke that presocialist past as their origin story that imagines their entrepreneurial activities as regaining China's glory and resituating China in a cosmopolitan world. Government officials involved in the fashion clothing import-export business admit the problems that existed in the socialist past but nevertheless they retain some sense of the importance of socialist goals, especially insofar as they are entangled with nationalist ones. Thus, as with elite entrepreneurs, these government officials explain their entrepreneurial search for profit as being linked to nationalist ideals, but insist they come out of the socialist past. Workers, meanwhile, blur the past with the present in their experiences of inequality, even as they do not wholly reject the socialist past as do elite entrepreneurs. They tell stories of the burdens their parents and grandparents have faced, even as they recognize continuities with their experiences in urban-based factories for export.

The second force underlying migrant workers' experiences of working in the transnational Italian fashion industry and the relationship of these experiences to historical transition is that of affective engagements. In the writings on affective labor, the emphasis is on affect as the product of one's labor or affective attachments to one's job, as fostered by corporate employers, especially in the so-called knowledge economy (Clough 2007; Hardt 1999). These approaches are important for linking affect with current capitalist practices. In my previous work (2007), I discussed the production of desire in China, as a key mode of understanding the experimental and unstable nature of what we might label, with some caveats, neoliberalism. In situations involving migrant workers in transnational Italian fashion production, affective engagements are principally not with the products of one's labor nor with the job itself but rather with temporality. The affective engagement I highlight here is with imagined futures. Again, these futures contrast with the affective engagements with temporality on the part of elite entrepreneurs, who engage with a revisionist history of the presocialist past, and on the part of government officials in this sector, who engage with the immediate socialist past.

Dorothy Solinger (2013) has also recently emphasized temporality, in contradistinction to spatiality, as an important measure of inequality in China.

Focusing on the vast dichotomy in material standards of living among urban residents—everything from food, leisure consumption, housing, education, health care, access to communications technology, and transportation—Solinger argues that the new urban elite have seen significant improvements in their lives while the urban poor have regressed to the poverty common among everyone in the Maoist era. Moreover, official discourse represents the former as signifying progress while the latter signify backwardness. Yan Hairong (2008) similarly argues that in official discourse migrant workers from the countryside represent a backward spatial dimension but also a forward-looking life when compared to urban workers, most of whom have been laid off.

My argument in this chapter builds on this work by focusing on how migrant workers grapple with these contradictory temporal representations. They do so by struggling to imagine themselves as part of China's future. Thus, the affective engagements I highlight here are engagements with imagined futures that allow migrant workers to view themselves as leaving the past behind. They include memories as well as yearnings. In the case of these migrant workers, the yearnings are for what is most hegemonic on the uncertain horizon of the future: that of becoming an entrepreneur. This dream of entrepreneurship is not just at the level of the social imaginary; it is also structured into the way in which labor in factories geared toward the export of fashion clothing is organized. That is, workers are treated as if they were already entrepreneurs of their own labor within the factories. This vision of themselves as entrepreneurs of their own labor displaces previous notions of class identification as a proletariat and partially disrupts a view of themselves as part of a working class in the present.

The transnational encounters that workers have with the Italian fashion industry are shaped by a prevalent colonial myth that continues to circulate in the West as well as in China: that China was an inward-looking, closed society prior to the West's ability to pry it open. This myth has been transposed to the postsocialist period and one often hears, both in China and elsewhere, that China was a closed country under socialism and only now is opening up to the world (Rofel 2012). Of course, this myth ignores not only the world of international socialism but also the fact that, throughout its long history, China has been engaged in several distinctive encounters that have turned into "worlding" projects, to borrow a term from Mei Zhan (2009). Thus, workers' encounters with Italian fashion production are posed as a new form of potential cosmopolitan knowledge, even as the colonial histories of textile production centered in Shanghai are erased through nostalgic histories for the presocialist past (for a discussion of these nostalgic histories, see Rofel 2014).

I use the term *encounter* to refer to engagements across difference. Attention to how culture making occurs through everyday encounters among members of two or more groups with different cultural backgrounds and histories and unequally positioned stakes in their relationships helps us to understand how "the cultural" is made and remade in everyday life. What we might call an "encounters approach" highlights how cultural meanings, identities, objects, and subjectivities emerge through unequal relationships involving people and things that might at first glance be understood as distinct. Such an approach also helps challenge an overly structuralist view of a transition from socialism to capitalism. Encounters prompt unexpected responses and improvised actions, as well as long-term negotiations with unforeseen outcomes (see Faier and Rofel 2014). Migrant workers' encounters with the Italian fashion industry are material and embodied. They include the actual fabric and garments made, with their particular sizes, designs, sewing, and quality demands. They also include the rhythm of work in a transnational export factory for fashion clothing, which is quite distinct from that of either production of large quantities of casual clothing or production for the domestic market.

Workers' ability to maneuver in the interstices between development and displacement is thus shaped not only by their encounters with the transnational supply chain for the Italian fashion industry and by their spatial migration but by a temporal movement as well. Their origin stories regarding the rural life they are trying to move beyond, coupled with their imagined relationship to a possible future of becoming something beyond being a worker, provide an unstable but potent means of leading them to endure their current exploitative working conditions. These three forces—the transnational supply chain for Italian fashion, origin stories, and affective temporal engagements—explain how migrant workers from heterogeneous life experiences embrace and produce new modes of being in historically contingent, sometimes surprising ways. Their articulation together shapes workers' experiences. That is, the dynamic engagement of these three forces points to the ways in which workers' transnational encounters of laboring in factories structurally shapes workers' own interpretations of the socialist past, which are their "origin stories" of where they have come from, even as their affective engagements with the temporality of both past and future in turn shape their interpretations of their experiences as factory workers.

This chapter is based on a collaborative project with Sylvia Yanagisako about transnational capitalism. We are examining what we call the twenty-first-century Silk Road between China and Italy in the fashion industry. Since

China's entry into the World Trade Organization (WTO) in 2001, nearly all the Italian firms that produce yarn, fabric, and garments for the fashion industry have moved their production to China. These production arrangements were originally all through joint ventures, as required by the Chinese government. Since China's entry into the WTO, however, the Italians have increasingly established wholly foreign-owned enterprises (WFOEs), which China had to permit as one of the conditions for joining the WTO, and which also have numerous subcontracting arrangements. The resulting twenty-first-century Silk Road runs from Milan to Shanghai. Shanghai and its environs are at the center of this particular set of transnational business relationships in China, because Shanghai is a large coastal city with the infrastructure to facilitate production and shipping, the textile industry has a long history there (including colonial relations of production), and foreign investors feel most comfortable living there.

The workers I discuss here worked in three different factories, each with a distinctive positioning in the transnational supply chain for Italian fashion. One is a large garment production factory in Tongxiang, an old canal town on the famous Grand Canal, located halfway between Shanghai and Hangzhou, which has been turned into an industrial zone. The Tongxiang factory was introduced to the Italian companies for which they produced garments through one of the Zhejiang provincial import-export corporations. The factory owner had met directly with the Italian customer, who wanted to see the conditions in the factories. The second, a yarn-spinning factory in Jiaxing, a city to the northeast of Hangzhou in Zhejiang Province, was originally a joint venture between an Italian company and both the Chinese central government and the Jiaxing municipal government. In the course of our research, the Italian owner bought out the Chinese government, and the factory became a WFOE. The third factory, located in Hangzhou, is the Zhenfu silk-weaving factory that I wrote about in *Other Modernities* (1999). Since the mid-1980s, Zhenfu had been privatized—bought by the former top cadre/manager—and now produces fabric for a wide variety of foreign customers. The factory receives its production orders directly from the Hangzhou Silk Corporation (formerly the Hangzhou Silk Bureau), under whose aegis they used to operate. This municipal corporation, which also operates as an import-export entity, makes all the arrangements without revealing the names of the customers to the factory. In essence, Zhenfu has not changed its relationship to the silk corporation except in one very important respect: With privatization, the government-owned corporation no longer takes responsibility for the wages or benefits of workers at the Zhenfu factory.

My concern on a broader scale is with how to analyze global capitalism and neoliberalism. Analyses of global capitalism and neoliberalism over the past fifteen years, such as those of David Harvey (2005), Immanuel Wallerstein (2006), and many others, have given us important insights into significant changes in the way capitalism operates. They help us to ask big questions that address systematicity. They are also fairly structuralist and at times abstract, and they assume an immanent logic to capital, as if capitalism contains ineluctable historical imperatives. Most important, they assume the existence of a separable field called the economic, thereby subscribing to one of capitalism's biggest fantasies: the division between the instrumental and the affective (Sahlins 1976). This chapter is part of my long-term efforts to offer an alternative approach to global capitalism that opens it up to cultural histories, (non-essentialized) subjectivities, and situated social action.

Origin Stories: The Temporal-Spatial Past

Xiao Lan is a thin, lively worker who was thirty-seven years old when I met her in 2007. She worked in the Tongxiang garment factory for export. As often happens with fieldwork, I fortuitously found my way into this factory by way of a friend of a friend. My connection at the factory, who later became a friend, is a young woman who was then finishing up her PhD on French views of China as reflected through cinema and literature. She held a managerial position, which she treated as temporary. When I arrived at the factory, we spent most of our conversation on her thesis and her various translations of well-known French authors into Chinese. While most factory owners and managers are leery of having foreigners view factory conditions or talk with workers, my friend threw open the doors for me to talk with as many workers as I wished.

Xiao Lan and I sat inside the Tongxiang factory offices and chatted about her life. She had grown up in Jinhua, another small town in Zhejiang Province about five hours from Tongxiang. Her parents had been farmers. Her mother died when she was young, and her father, when she was eighteen. Her parents had lived in poverty and sometimes hunger. Xiao Lan came of age at the beginning of economic reform, when rural collectives were broken up, economic production devolved to the household, and rural residents could enter into commerce. Xiao Lan's father had opened a small store in Jinhua.

Xiao Lan has lived a life of constant movement, a narrative I heard from many workers. She, like these others, is considered a migrant worker. Xiao Lan had first worked in her hometown of Jinhua, in a factory that made clothing for export. She worked there for six years, but only on and off because, as she described herself, she is someone who likes to "play" (*wan*). She then moved from her hometown, to the slightly larger town of Huzhou, also in Zhejiang, where she worked for a year in a factory that made high-fashion women's clothing for export to Italy, Africa, and the central Asian countries surrounding China. She had been encouraged to go to Huzhou by the factory owner, who was from her hometown. Finally, the same year I met her, she moved to Tongxiang to work in this factory. Tongxiang had recently been formally recognized as a city but it is still small compared to the megalopolis of Shanghai. In each of these towns, Xiao Lan had worked in garment factories for export. I asked how she had learned to make clothing. She said she had studied with a *shifu* (master craftsperson) for two years. That *shifu* had her own store. She learned how to sew, cut the material, measure people—everything. The *shifu* didn't pay her but she helped the *shifu* with her business for two years. By the time she moved to Tongxiang, Xiao Lan had been promoted to quality inspector. She walks around and checks on people's work to correct mistakes.

Xiao Lan describes herself as someone who, in addition to playing around, is also brave. By bravery, she meant she never worried about moving around from place to place and job to job. What struck me about Xiao Lan's movements, however, was not their spatial dimension, but their ontological instability. Only when I asked about her plans for the future did Xiao Lan explain that in between working in the Jinhua factory, she had opened a restaurant and then a dance hall. She had owned the restaurant for three years, and she really likes to cook. But then the owner of that building tore it down and sold the land to someone else to build a high rise. She did not make a lot of money from the restaurant, but she enjoyed it. Then she opened a dance hall with her boyfriend, but they fought so she left. She decided to engage in manual labor (*dagong*) because it was less hassle (*mei fannao*). She would like to think about opening a restaurant again, this time in Tongxiang. Thus, she has moved back and forth between working in factories and being a small entrepreneur.

Xiao Lan's story alerts us to a set of origin stories, affective engagements, and transnational encounters among workers that are quite distinct from those of the elite entrepreneurs I had come to know within the supply chain for Italian fashion (see Rofel and Yanagisako forthcoming). Xiao Lan and other migrant workers I met do not engage in a fierce condemnation of the socialist past, as do those elite entrepreneurs, but neither do they embrace it. Nor do they draw a strong divide between the past and present and castigate the supposed lingering effects of socialism in the present, as do those elite entrepreneurs. Xiao Lan's and other migrants' origin stories reflect a much more fluid and ambiguous sense of both continuity with and change from the socialist past. As Gail Hershatter has alerted us in *The Gender of Memory* (2011), regarding rural women's lives under Maoist socialism in China, the oral histories of China's rural citizens do not subvert so much as transverse official histories.

The origin stories of the migrant workers whom I met are about moving away from their parents' and grandparents' memories of poverty under Maoist socialism, a poverty they experienced well into the post-Mao era. Moreover, they had moved from an ontology of being a peasant, which has, in this new era, come to signify backwardness and immobility (Anagnost 1997). Whereas once the socialist state secured life for people in the countryside, now households must attain security on their own. This more insecure life also means more mobility for these young migrants, which in turn signals both possibilities as well as hardships.

One of the main reasons for the ambiguous sense of continuity with, as well as change from, the socialist past is the *hukou* system, a system of residency instituted by the socialist state in the mid-1950s that created a two-tier citizenship. Every person receives a residency permit, and until recently someone could only live in the area specified on that permit. This division had enormous consequences under socialism, because only urban residents got access to full social benefits and, importantly, food. The socialist state used this *hukou* system to pursue its developmentalist policies of urban industrialization, taking resources from the countryside to support the cities. This division lives on into the present, in that rural migrant workers can get temporary permits to work in cities, but they do not receive any social welfare. Moreover, urban residents disdain rural migrants, accusing them of being responsible for the crime, dirt, and other social ills that plague the cities (Zhang 2001).

While the rural/urban division is prominent for these migrant workers, as an overarching sense of division among citizens in China, these particular migrant workers have actually moved around on a circuit that, for the most part, did not include the major eastern coastal cities of Beijing and Shanghai. Instead, they circulated through small towns in Zhejiang Province—such as Jinhua, Huzhou, Jiaxing, Tongxiang, and Wenzhou. Their discomfort in some of these towns had more to do with what they described as native-place prejudice. Xiao Lan, for example, said she was not comfortable in Huzhou and that she didn't like working there because Huzhou people were not very nice. They are jealous (*duji*). I asked what she meant. She said that when you did well, they would be full of jealousy and if you made a mistake they would laugh at you. I asked why, and she said it was because the people are *bendi* (locals) and they are this way with the workers who come from elsewhere (i.e., migrants). They don't give people a feeling of accomplishment (*chengjiu gan*).

This division between rural and urban citizens is one of the key reasons that the socialist past and the postsocialist present have a more fluid and ambiguous temporal relationship for these migrant workers. If they are leaving the past behind, it is a past that spans the division. Unlike elite entrepreneurs, who insist the past should stay as the past though it frequently refuses to do so, for these migrant workers, it is a past whose elements, including most importantly social relations of inequality between urban and rural citizens, seamlessly blend pieces of socialism with the aftermath of socialism proper. These socialist and postsocialist elements coalesce for these migrant workers in a manner that puts the grand narratives of transition off center. They displace, as well, elites' stories about the origins of social inequality, implicating urbanites more directly in their interpretations of their experiences as migrant workers. The socialist past was certainly a contradictory endeavor of ideological commitments to equality along with a drive to achieve wealth and power in a modernist, developmentalist mode (Lin 2006). This latter technocratic mode, combated by Maoism and utopian socialist thought, certainly became triumphant in the Reform Era. Current inequality is thus a continuation of prior developmentalist practices combined with novel forms bred out of the specific mode of capitalist accumulation China has embraced.

Affective Engagements with the Future

Xiao He was a young man around thirty years old when I met him in Tongxiang. He learned how to sew clothing when he was a little boy in Wenzhou, a city in southern Zhejiang Province. I pressed him on this question of how he had learned to sew, since one aspect of the factories that struck me was the re-entry of men into textile production. Prior to the socialist revolution, large-scale textile factories were found mainly in Shanghai, where workers were women. By contrast, in the Hangzhou area, textile production, especially of silk, was largely in household workshops, where men did the weaving while women prepared the thread. Once the socialist government pulled these household workshops into large-scale factories in the 1950s, they instituted a gendered division of labor that increasingly made silk weaving into women's work. By the 1980s, when I had conducted my research on the silk industry in Hangzhou, there were no men—none whatsoever—doing either the thread production or the weaving. Men fixed machines, did the dyeing, and, of course, were managers/cadres. But in 2007 when I went into the textile factories for international export, I found that approximately 50 percent of the weavers and sewers were men. Later in the afternoon, as I walked around the factory shop floor with Xiao He, I asked him why he thought men are willing to do this work, and he replied, "We used to *zhongnan qingnü* [look up to men and look down on women]. That was in traditional society. But now people are freer, and they can make a choice." I might also infer that textile production labor is less onerous and less dangerous than construction work or other types of factory work, and therefore that has attracted men.

Xiao He responded to my question about learning to sew by expanding his narrative about his boyhood: When he was a little boy, he had an interest in sewing and occasionally, when he had nothing to do, he would sew a little. I asked when he started to work. His family was poor and he had a lot of brothers and sisters, so he was apprenticed out at a young age to a *shifu*, who had a small store, in order to learn how to make clothing. Xiao He spent one year at the store with that *shifu*, and together they made ordinary clothing. Then he spent another year in Shanghai with another *shifu*, whom he met as a result of introductions by others. With the latter *shifu* he learned how to make Western clothing (*xizhuang*). I asked if there was any difference in learning how to make Western clothing, and he said, yes, it was hard to learn. I asked why, and he explained you have to sew by hand, but he was determined to learn (*qiangbo yizhi*). Then he went to Shanghai for a year and apprenticed to yet another *shifu*. After that, he worked in Wenzhou for ten years in a factory that made Western men's suits for an Italian label as well as two other labels, one of which was for the domestic market.

Xiao He rose to become head of one of the workshops. He said that it was challenging because you have to figure out what workers are thinking and they are not always happy (*xin buping*—literally, their hearts are not calm). I asked how he figured out how to do the job. He said he just groped around (*mohe*) for a way to do it. One issue workers were unhappy with was freedom (*ziyou*). I believe that by this he meant workers being able to move around at will in the factory and to come and go as they pleased. We could call these micro-temporalities or quotidian temporalities. Of course, there are factory rules (*changgui*), he said, so there was nothing he could do to resolve this problem. They have to follow the rules. The other issue was making money. They want to make more money, and he could do something about this. He told them if they raised the amount they were able to produce, then he could give them more money, and they were satisfied with that. He left that job and moved to Tongxiang to be with his wife, who had moved there two years ago to work in a garment factory.

I asked if he had plans for the future, and he said he would like to stay in Tongxiang, adding that he doesn't like to move around. He noted that some workers like to move around a lot, and they keep changing jobs to look for better wages. But they are always unhappy. Really, he added, the wages are all about the same; the problem lies in their hearts (*xinli wenti*). I asked if he had any other plans for the future. In response, he began by explaining he did not have any grand ideals and just wanted to lead a decent life (*ba shenghuo gaohao*). But then he added that he used to have his own shop. It was a small workshop where he worked for himself and had a few others he hired. They made clothes for others (*jiagong*), meaning that when factories have an order they need to fill and are under pressure, they turn to these small workshops to complete the work. But it didn't work out. He would like to think about doing it again, but it is hard to save the money for that. And it is not always successful, like this factory that went bankrupt before (meaning the factory he is working in now). Of course, if you are someone with a lot of money, then it doesn't matter.

Unlike elite entrepreneurs in Shanghai, these migrant workers do not feel nostalgic for a presocialist past or for any other past. Their affective engagements are not with the past but with a fantasy of a future, one in which their ontological status will not be that of a peasant, or that of a migrant worker, but rather that of someone making a successful business on his or her own. I have often been asked why these workers put up with such horrendous working conditions. Of course, one could go a long way by answering that they have no choice. While that answer is important, it is incomplete. Workers are quite articulate about their horrible working conditions. Contributors to Litzinger's SAQ special issue (2013), for instance, have pointed out that China's new generation of workers are better educated and therefore can give voice to a range of insights about their living and laboring conditions.

Similarly, at the Tongxiang factory, workers described their working conditions to me very clearly: They get paid by the piece, and when there is work they have to work continuously, with no days off. Sometimes they even sleep by their machines, because they work overtime in twenty-four-hour cycles to finish the order for export. When there is no work, however, they get sent home with no pay. They get about two days off from work each month, and they often work from eight in the morning until nine at night. They work as many hours as they need to when they have an order to finish and then they get some rest time. So they don't have a regular work week and don't have regular days off. There is no guaranteed wage in privately owned factories for export. These working conditions in a garment factory geared toward Italian fashion export are somewhat distinct from the Jiaxing spinning factory or Hangzhou's Zhenfu fabric-weaving factory, though both are also geared toward export. These two factories are based on the old socialist work schedule, the three-shift system: two days of day shift, two days of evening shift, and two days of overnight shift. Then workers have two days of rest. In the 1980s, workers used to explain how difficult this shift system was—from the difficulties for their bodily rhythms and struggles with sleep, to the challenges of arranging their family lives, including childrearing. In the current neoliberal moment of commodity-chain production, workers regard this shift system as relatively benign compared to the situation at garment production factories for export. Yet in all the factories, the inequalities the workers face in their daily work experiences are structurally visible to them, as many migrant workers have worked in three, four, or five factories. They therefore do not blame individual factory owners but the whole factory regime.

Ching Kwan Lee (2007) describes labor protests in southern Guangdong Province by migrant workers in production factories for export. They protest unpaid wages and illegal wage rates, disciplinary violence and humiliation, and industrial injuries. They turn largely to the legal system in hopes of redress from the state. But they also engage in street protests. Pun Ngai (2005), who also did research in Guangdong, points to more submerged but nonetheless evident forms of "transgression," as she calls them, in which incipient subaltern resistance has the potential to form. These transgressions include workers speaking to one another about their working conditions, daring to voice criticisms, or covering for one another at work. Pun and Jenny Chan (2013) more recently have documented the activism of workers in transnational supply-chain factories, such as sharing organizing skills and disseminating protest strategies, bringing specific issues to management, and presenting demands to the local government. They emphasize the central role of what they call the "dormitory labor regime," in which factory owners house and feed workers at minimal levels in order to better exploit their labor. This dormitory labor regime fosters the discontents—as well as organizing abilities—that fuel workers' protests. Ho-fung Hung further argues that the protests by migrant workers in the export-oriented sector have also played a role in helping the central government pressure local governments to accept labor reform (2013: 210).

In the lower Yangzi River Delta area where I conducted this research, migrant workers mainly protested by leaving one factory for another. Indeed, one of the factory owners' and managers' main complaints to me about production for export concerned the instability of their workforce. For instance, when migrant workers return to their hometowns and villages for the Lunar New Year holiday, factories would typically lose about 20 percent of their workforce because workers would fail to return. Several of these factory owners have turned away from production for export and instead focus on the domestic market, so that they can offer more regularized working conditions and thus have less turnover.

While workers are quite clear about the depredations of their working lives, they move through this world in the hope of moving beyond it. While they have a clear critique of their labor conditions, and often vote with their feet by leaving one factory after another, they nevertheless refuse to embrace the ontology of being a worker. Under socialism, workers were touted as heroes and indeed at least for urban workers this official display of adulation did change the meaning of their demeaning factory work.

But migrant workers from the countryside have never been celebrated in this way. In the post-Mao era, there is no glory in being a worker. On the contrary, workers are placed on the lowest rung of society, sometimes to be pitied, sometimes to receive some protection from the state, but mostly as those from whom others distance themselves and with whom they no longer see themselves as having any relationship. Thus, many workers included in their narratives a story about how they planned to move beyond working in a factory. Each and every worker told me how they used to have, currently have, or will have their own workshop or shop. Some had small garment workshops like Xiao He, supplementing factory needs. But others, like Xiao Lan, opened or dreamed of opening restaurants or flower shops.

Tang Qing, from Sichuan Province, was twenty-three years old when I met her in 2008. She worked in the Jiaxing spinning factory, and by that point she had worked in several different kinds of factories for seven years. Her parents, she said, were peasants (*nongmin*) whose life was very bitter. They had only three *mu* of land (1 *mu* = 1/6 acre), on which they planted rice, wheat, and occasionally corn. These are staple crops that do not bring in a lot of income. Tang Qing helps them out with the farm work when she is home, and she also sends them as much money as she can from her wages. She would have liked to continue in school but her parents were too poor to pay the school fees, so she began factory work instead.

> TANG: I worked in a factory near my family home (*laojia*) for one or two years. But they asked for a big deposit, it was too expensive, they asked for ¥1,000–2,000. Other factories were too far away from home. That factory I worked in made plastic products. When I first came to Jiaxing, I worked in a shoe factory. I worked a little more than one year. But we had to work overtime all the time and the wages were not that high. It was too tiring. Then I worked in a candle factory for a year. Then I decided to go home. I was not planning to come back here. But I couldn't find work at home, so I returned.[1]
>
> ROFEL: How did you come to work in this factory?
>
> TANG: I heard this factory was not too bad, but at that time they wanted only local (*bendi*) people. But then they started to accept people from "outside" (*waidi*). I've been here more than one year in this factory. The work is not as tiring and we don't have to work as long. We don't have to work overtime. The wages are not so bad, but the prices of everything have gone up a lot.

Tang Qing is not yet married. She and her boyfriend plan to open a flower shop in his hometown in Jiangsu Province, once they save up enough money.

This affective engagement with a possible future in which one could have another mode of being is what pulls migrant workers through the present. Ching Kwan Lee (2007) has similarly argued that migrant workers want to leave behind their "vivid memories of and aversion to poverty associated with collectivized agriculture." Moreover, she argues, "bereft of an alternative vision of the social order," migrant workers in Guangdong, like migrant workers in the lower Yangzi Delta, "aspire to entrepreneurialism" (2007: 206).

This dream of becoming an entrepreneur has been fostered for over a decade in both official and popular narratives. Soap operas, journalistic stories lauding successful wealthy businessmen (and they usually are men), autobiographic success stories by some of China's *Fortune* 500, and the Chinese Communist Party's decision to allow entrepreneurs to become Party members have conjoined to make the dream of becoming an entrepreneur the main road to self-development in this neoliberal moment.[2]

Becoming an entrepreneur is the dream encouraged by official and popular discourse alike. Numerous soap operas are full of heroes who are entrepreneurs, and who embody the figure of the morally upright, nationalistic citizen who used to be represented by Party cadres or People's Liberation Army fighters. But these latter figures now represent corruption, and entrepreneurs have taken their place. The Communist Party has also put its official imprimatur on entrepreneurs, admitting them to membership in the Party as representatives of "the people." Indeed, there seems to be no other dream on the horizon at the moment. When I visited China in December 2013, even former workers at Zhenfu, who had worked under the old socialist system, were all scrambling to open their own small businesses. The former head of the Youth League, for example, had opened a restaurant. Another former worker in the spinning workshop had opened a real estate business.

This alternative mode of being, however, is not just the *homo economicus* of classical economics, nor the one being lauded by the Chinese state. It is a mode of being in which migrant workers' lives and their very subjectivities are meaningful and socially validated. Thus, the quality of their desire to be moving into another way of life exceeds their existing mode of being. I have argued elsewhere that the socialist narrative of speaking bitterness engendered a desire among subalterns for a "historical imagination of overcoming" (1999: 141)—in that case, overcoming the preliberation past by embracing the socialist present. Migrant workers have developed another historical imagination of overcoming, in this case, the (harsh) present by embracing a possible future. Yan Hairong calls this an "ephemeral transcendence" (2008: 8), as most migrant workers will never be able to become entrepreneurs. In this sense, their futuristic aspirations resemble those who hope to migrate elsewhere—especially to Europe or North America. As Julie Chu (2010) has so eloquently shown, those from China who hope to migrate to these elsewheres often construe their preparations for such often unrealized dreams as in themselves a form of mobility. Here we find a very close resonance with the migrant workers I met in the factories. For their various preparations—fantasizing possible entrepreneurial paths, saving for that dream, enlisting others to dream with them—resemble these rural Fujian residents Chu writes about.

There was an important exception to this future-oriented affect centered on entrepreneurship: older workers who were married and had a child. For example, Luo Ming was a thirty-five-year-old spinner at the Italian-owned Jiaxing spinning factory when I met her in 2008. She was from a nearby township called Wanjiajing. Her father farmed and her mother was a sent-down youth from Shanghai, who had stayed and was still working in a factory in Wanjiajing. Luo Ming had been working in this spinning factory for eleven years. She was not very good in school and didn't like just sitting at home, so she entered the factory. Some of her friends had opened their own businesses, but they had not done that well and had also entered the factory. Luo Ming had just signed a contract to become a permanent worker in the factory. When you get to be forty years old, she said, it's difficult to find work. She understood that she could retire at the age of fifty. Her husband also worked in a factory, one that made car parts. Luo Ming pinned all her future hopes on her daughter, for whom she spent quite a bit of her wages to send her to the Youth Palace to learn how to study harder and to play the piano. She looked forward to her retirement. Her vision of the future was not for herself, but rather for her daughter.

Transnational Encounters

Xiao Zhan worked for four years in a leather clothing factory in the nearby town of Chongfu, his hometown, before coming to the factory in Tongxiang. He had slowly moved his way up to the job of quality inspector. His family had lost their land to the local government for real estate development and his father had subsequently opened a small shop to sell dry goods. Before working in the leather factory, Xiao Zhan had opened his own small workshop, making clothing. He didn't need any capital to open it and had learned how to make clothing from his older sister and his older aunt's daughter. His older sister works in a clothing factory in Haining (a small town north of Hangzhou, in Zhejiang Province), and his mother was a *shifu* in a nearby clothing factory. His workshop did well and he hired others to work with him. He ran the workshop for three years before going to work in a factory. Indeed, he still has the workshop, and his father helps to look after it. He plans to expand it in the future. It's hard work to run the workshop but very stable. He only went out to do factory work, he said, because his mother wanted him to get some experience out in the world. He was twenty-eight years old when I met him.

Xiao Zhan, as with all the workers with whom I spoke, makes a clear distinction between working on fashion clothing for export versus working on ordinary clothing for the domestic market. The former is more difficult, has a much more onerous temporality, and the pay is less certain. Xiao Zhan said that with ordinary clothing, you are required to produce a large quantity but it does not require a high level of skill, while with high-fashion clothing the quality is high and consequently you have to really make an effort. Xiao Zhan observed that what makes fashion clothing for export more difficult is how much close attention to detail workers must give—how the garment is sewn, how the seam around the shoulder is put together, making sure that all the seams are straight. Xiao He also echoed this point, noting that you can't tear out the seam once the whole garment has been sewn. Hence, the importance of the job of constant inspection. Both Xiao Zhan and Xiao He emphasized that women's clothing for export was more difficult than men's. The frequent changes in the styles of women's clothing require workers to continually learn how to sew new garments. Xiao He's view was that there should be a difference in pay between working on fashion clothing for export versus working on what he and others called "ordinary clothing."

With fashion clothing for export, time schedules must accommodate the foreign customers' last-minute demands. Xiao Zhan stated unequivocally that the main problem he faces with this work is overtime, which is usually obligatory. Workers often do overtime because the foreign customer has set a deadline and sometimes the factory has only twenty-four hours to meet it, because otherwise it would be charged a penalty. Often, foreign customers only request a comparatively conservative order, and once it sells out they need a supplementary order on short notice. Alternatively, they may make last-minute changes to the design of the clothing. On those days, workers labor for twenty-four hours straight, sleeping intermittently near their machines. As Xiao Zhan averred, the foreign customer is the one who has the sole power to decide (*kehu shuole suan*). Thus, the boundary between work and nonwork time has increasingly blurred. Indeed, the structure of labor in factories for export has required of workers that they treat their labor in an entrepreneurial manner. That is, they themselves must assume the risks entailed by the factory's ability to make a profit. This creates not only the sheer fact of employment and possible layoffs, but also the belief that workers must be willing to do whatever it takes to create a commodity for export. Thus, workers go through "booms and busts" in these factories, in which they either work intensively throughout a specific time period or have no work. In the latter periods, they also have no pay. They are "on call" to the factory owners at a moment's notice and must find other means of survival in the interim but have to be ready to drop other arrangements when they are called in to work.

I asked Xiao Zhan which he would prefer to work on: fashion clothing for export or ordinary clothing. Most workers responded that they would prefer to work on ordinary clothing, because then they don't have to think so much about what they are doing. Xiao Lan, for example, said that working on fashion clothing is a lot harder because the styles change so often. You have to look more closely at what you are doing and pay attention all the time. With ordinary clothing, once you get the style, you can do it easily because you keep doing the same thing over and over again. She prefers to work on ordinary clothing, noting that it is more comfortable to work on a large amount of ordinary clothing, and you don't have to think too much about what you are doing. Xiao Zhan, by contrast, said he will work on whatever gives him food to eat, and he concluded that if you can't "eat bitterness" (*chiku*), then you can't do this work—ironically using the primary phrase that the Community Party had employed not only in its early organizing but also after it established the PRC, to describe the condition from which they had liberated China's masses.

The workers' transnational encounters are vivid, but are of a different mode than those of elite entrepreneurs in Shanghai. Their encounters are with the materiality of what they produce, which links China to the Italian fashion industry. Their encounters are with transnational garments, but also constitute a distinct means of encountering the foreign as well (Rutherford 2003). The fact that Xiao Zhan's mother wanted him to "see the world" by going out to work in a garment factory for export alerts us to the importance of these encounters for migrant workers.

Because these migrant workers are also treated by urban citizens as though they are virtual foreigners, it is not entirely unreasonable to state that their transnational encounters also include their process of coming to live in China's cities. In the cities, these rural migrants learn a range of different forms of cosmopolitan knowledge they hope will help them become entrepreneurs, including various forms of consumption and leisure activities (Pun 2003; Yan 2008), clothing styles, housing arrangements, as well as the temporality of labor and various ways of making a living. Even in the more regional circuits that these migrant workers follow, they nonetheless learn this cosmopolitan knowledge. Often, they are moving to larger towns or cities than the ones they came from. Yan Hairong (2008) has shown how urban consumption, which signifies a modern subjectivity, is an important source of attraction for young rural women drawn to Beijing. She calls this a "mirage of modernity" (145). Pun Ngai has described how consumption is a new "technology of the self" for migrant women who desire to become modern urban subjects (2005: 157–163). While Yan insists that consumption does not resolve the contradictions that these women face as marginalized workers from the countryside, Pun argues that consumption leads women away from a critique of the conditions of their exploitation.

Migrant workers also experience new modalities of inequality. Whereas in their home villages inequalities are centered on local officials, in Tongxiang, Jiaxing, Hangzhou, and other eastern coastal cities, these workers find themselves subsumed into the overarching category of the migrant. It is partially the experience of being "other" to urban and local citizens that blurs for these workers what urban elites take to be clear historical divides—or what *should* be clear historical divides—and thus shapes their interpretations of current social relations of inequality. The workers I have described here, however, do not enter into the megalopolises of Beijing and Shanghai. Their circuit is within the smaller cities in Zhejiang Province, and therefore they still retain a regular connection to their hometowns; some still live at home. Others, when they described the disdain they experienced away from home, attributed it as much to the small-town native-place thinking of the citizens of Jiaxing or Hangzhou as to an "urban" experience.

Whereas elite entrepreneurs also cast transnational entrepreneurs in the lead role of exploiter, migrant workers link the conditions of working in an export-oriented factory to the domestic factory regime. The grueling conditions of factory work, urban and local citizens' disdain, and working for export within a local factory together shape the conditions of hardship that workers dream of leaving behind. Yan Hairong (2008) further emphasizes what she calls the "emaciation" of the rural, by which she means that the countryside has been eviscerated of state investment, sociality, people, and even production, as more and more people are encouraged to migrate to urban areas. Starting with Hu Jintao and continuing into the present, the central state leadership has begun to encourage more investment in rural areas, but their goal is to urbanize much of the countryside.

Thus, to understand workers' experiences of social inequality in China today, it is important to pay attention to the specificity of their transnational engagements. While migrant workers across different kinds of industries certainly share overarching problems of exploitative and degrading working conditions, it also makes a difference exactly which kinds of working conditions they face, including the specificity of the sector in which migrant workers labor, the labor conditions in the factories of that sector, as well as the materiality of the commodities they produce.

Conclusion

I began this chapter with two interlocking questions about how people move from one world to another and the role of transnational commodity chains in shaping migrant workers' experiences of that transition, through working in factories for export. My answer was to show how the transnational aspect of their labor has an intertwined relationship with two other modes. These three modes are (1) origin stories, by which I mean stories people tell about where they have come from and how that has shaped their current situation, including social inequalities; (2) affective engagements, by which I mean the attachments and embodied experiences people form in relation to their labor and to social relationships by way of history and memory; and (3) transnational encounters, or worlding projects, by which people construct and get pulled into a certain vision of what constitutes a meaningful world.

Unlike the elite entrepreneurs I came to know in the transnational commodity chain for the Italian fashion industry—who vehemently reject the socialist past even as they feel that past continues to live on in the present, and in whose view there should be a sharp division between then and now even as they are affectively engaged in a nostalgia for the presocialist past—migrant workers do not renounce the socialist past, for the simple reason that the boundary between that past and the present is not a sharp one for them. This is not to discount the real differences that exist between life under socialism and under postsocialism for those in the countryside (see Yan 2008). However, it is also important to acknowledge that, from the beginning, the socialist state had extracted resources from the countryside to focus on urban industrial development and had offered full socialist welfare only to those living in the cities. The *hukou* system implemented under socialism hardened these differentiations.

Migrant workers' origin stories come out of this blurred past and present. These origin stories move migrant workers, affectively, toward an imagined, possible future of becoming something else, the one thing that seems visible on the horizon—that of becoming an entrepreneur. Again, in contrast to elite entrepreneurs, who ironically resemble Hegelian Marxists in their belief that temporality and history should be properly linear and progressive and have a telos, these migrant workers could be seen as combining a certain Maoist approach to temporality—skipping among historical periods—with a Benjaminian view of the ruins of the past piling up at their feet. They do not embrace a telos as they face the precariousness of their futures, even as they hold out a measure of hope.

Affective engagements with temporalities, with the past, present, and future, and with memories and revisionist histories, are what compel elites and subalterns alike to reenvision the worlds they are both pulled into and create. They include embodied sensibilities that are not merely located within the individual body but also are produced in relation to others.

The resulting transnational encounters, or their worlding projects, meanwhile, entail material encounters with the products of their labor as well as the temporalities of foreign companies. They also include living among China's urban and local citizens, as well as the latter's disdain for these migrants. Migrant workers' interpretations of social inequalities in China thus meld together precisely what elite urban entrepreneurs keep apart—that is, the production of social inequalities through capitalism. It is precisely the commodity-chain nature of their labor that makes these encounters formative in shaping how migrant workers dream about their futures. Their awareness of the various elsewheres that have changed life in China, their material involvement in the garment production, their understandings of the challenges in this labor that gives them a modicum of pride in their ability to do it, their work rhythms, and their encounters with others within the factories—all of these shape their senses of the meaning of their own lives. Most important, for this generation of migrant workers, the distinctiveness of their lives when compared to those of their parents is thoroughly intertwined with that of transnational capitalism.

Attention to encounters bridges the division between structural histories of socialism and capitalism, and the meanings people give to their labor and to the people, resources, and objects they encounter and produce. Such an approach helps us to see contingency, unexpected outcomes, and articulations of multiple practices that make capitalism an ongoing process of creation and destruction rather than a singular, deterministic structure. It attends closely to the dynamic production of inequality across space and time. Attention to encounters thus offers a means for systematically examining political economic processes by recognizing the forms of difference that constitute them.

The role that these transnational commodity chains play in the lives of migrant workers encourages us to reverse the way we approach transnational capitalism and neoliberalism, treating them not as external *context* for local enactments or as sui generis phenomena but rather as the effects of social engagements that produce their nondeterministic systematicity. Attention to the articulation of origin stories, affective temporalities, and transnational encounters helps us to develop this nondeterministic approach to capitalism—one that accounts for how people, in their daily lives, grapple with, maneuver among, take in unpredictable directions, and occasionally subvert dominant socioeconomic and cultural orders. How these elite entrepreneurs and migrant workers cum entrepreneurs both shape and are shaped by the worlds they move within: This is an ongoing theoretical question we should continue to examine. It is also a question that I feel must be addressed if we are to search for the worlds that are more socially just.

Notes

1 Yan (2008) and Pun and Chan (2013) have pointed out how the production of labor in China depends on the social reproduction of migrant workers' return in the rural areas from which they come.
2 See Yan (2008) for a description of how domestic servants who have started their own businesses are lauded for their success at self-development.

9

REGIMES OF EXCLUSION AND INCLUSION

Migrant Labor, Education, and Contested Futurities

Ralph A. Litzinger

The earthquake that devastated central Sichuan Province on May 12, 2008, left nearly a hundred thousand people dead and millions more homeless. Beyond its immediate impact, the quake also provided a catalyst for an array of structural transformations that had far-reaching consequences in their own right. Entire towns and cities were leveled, and new communities were often quickly constructed on the sites of the destroyed communities. While in some cases these new constructions simply represented an extension of development plans that had already been in place before the catastrophe, in other cases, as Robin Visser discusses in her chapter in this volume, municipal governments and urban planners took the opportunity to radically reconfigure the region itself, taking land that had formerly been used for public services and reappropriating it for corporate use.

One of the controversies that emerged following the earthquake involved the thousands of children who were crushed when their schools collapsed—a result, many contended, of developers having cut corners by constructing the school buildings out of cheap, substandard materials, resulting in the now-notorious "tofu-dregs construction." While the Chinese government and the official press initially attempted to downplay this aspect of the tragedy, a vigorous grass-roots campaign (spearheaded by the Beijing-based artist and social activist Ai Weiwei) worked for months to identify each of the schoolchildren who died in the quake and publicize their names. Eventually, after refusing for months to address the issue, the Chinese government finally announced—on

May 5, 2009, just days before the one-year anniversary of the quake—that 5,335 children had died in the disaster.[1] A key issue raised by these efforts is not only the attempts to memorialize the children who perished and to remedy the structural problems that may have contributed to their deaths, but also the ways in which different groups were attempting to use the children's names and corresponding statistics for differential ends. Who decides whether or not information on the dead will be publicized, and who shapes the corresponding public narratives? Who is entitled to speak for the deceased and to decide what would be most advantageous for other children like them?

Like the earthquake itself, the grass-roots campaign on behalf of the dead schoolchildren had long-ranging ramifications of its own. For instance, when school buildings in Beijing were later inspected to make sure that they were all up to code, the schools that were most directly impacted were neither the public schools that were owned and operated by the city nor the expensive private schools that had considerable financial resources at their disposal, but rather the dozens of independent schools that had been established in recent years to meet the educational needs of the children of migrant workers. Because public schooling in China is only guaranteed to families with a local *hukou*, it is generally not available to children who accompany their parents into the city for work (Kwong 2004). Moreover, many private schools in Beijing and other Chinese metropolises are quite expensive and out of reach for the vast majority of migrant families. As a result, a number of independent schools have been established in Beijing and other major cities to meet the needs of this vulnerable demographic.[2] These schools are open to nonresidents, and they either charge a significantly lower tuition than ordinary private schools or offer full-tuition fellowships. Because many of these schools simply used whatever buildings were available and furthermore were operating on a shoe-string budget, they sometimes encountered difficulties in satisfying the postquake building inspections. The irony, accordingly, is that the postquake attempts to inspect school buildings in order to secure children's safety ended up disproportionately affecting precisely the cohort of migrant children whose position was most precarious to begin with.

One of the Beijing schools for migrant children that was threatened with closure by these postquake developments was the Dandelion School, where I have worked for the past seven years.[3] Located in the southern part of Beijing near the city's Nanyuan Airport, the Dandelion School is a not-for-profit institution that was established in 2005 and is largely funded by private donations. It is expressly designed to serve the children of migrant parents who come to

Beijing to find work. Currently serving more than six hundred children, the Dandelion School is one of Beijing's longest-running migrant schools.

Like many of its peers, the Dandelion School has attracted support from various corporations and philanthropies, scholars and activists, and students and researchers, and the result has generally been mutually beneficial. The migrant children are given educational possibilities they otherwise probably would not have had, the children's parents are able to earn a higher wage in the city than they would have otherwise, the philanthropists are able to contribute to a meritorious cause, while the scholars and students obtain material they may use in their research. At the same time, these sorts of synergies may also generate significant conflicts of interest, as different agents negotiate their objectives with those of the children themselves. One of the key questions, however, is what in fact are the best interests of the children, and who should be permitted to speak and act on their behalf. The children themselves are not passive or silent interlocutors in these debates, and their needs and desires are not necessarily aligned with those of the adults around them.

Orna Naftali (2014) argues in *Children, Rights, and Modernity in China* that there has recently been a discursive shift in China toward viewing children—and particularly urban children—as "rights-bearing, autonomous subjects." Although Naftali focuses primarily on middle-class children, her analysis nevertheless has interesting implications for the children of migrant workers, who are themselves perceived as lacking many of the basic rights that residents with an urban *hukou* take for granted. As a result, these migrant children find themselves at a peculiar chiasmus, wherein as children they are increasingly perceived as rights-bearing subjects even while, as migrants, they are systematically denied many of the social rights that locals can ordinarily rely on.

More generally, Eli Friedman (2013; see also 2014) argues that the state, in creating a vast underclass of migrant workers in the cities where they work, has managed to "magically" separate economic production from social production. He observes that China's mobile, floating, rural migrants provide strong bodies, cheap labor, and a lot of desire, yet they cost China's cities almost nothing. Instead, the production of migrant labor power itself—what Friedman calls social production—takes place almost exclusively in the countryside. In effect, the social and financial costs of education, health care, housing, and dependent care are all supposed to happen "back home," in the countryside. In practice, however, the costs associated with this process of "social reproduction" of migrant labor are not borne entirely by the rural communities, and a pressing question facing contemporary China involves how to redistribute these associated rights and responsibilities as many migrants and their families are developing stronger attachments to the cities to which they have come to work.

To address these issues, I will consider three recent documentaries that each focus on the children of migrant workers, specifically on the question of their access to education.[4] Some of the children featured in these documentaries have come with their parents to the city, while others remain behind in the countryside while their parents work in factories to support their education. The resulting films, meanwhile, dramatize conflicts between the parents' objectives and their children's desires, while also tacitly raising the possibility that the filmmakers' interests may diverge from those of the families they are documenting. In this way, the documentaries examine and problematize issues of education, futurity, and the ethics of observation and representation, as they coalesce around the figure of the migrant child in contemporary China.

Nowhere to Call Home

The most recent of these three documentaries is Jocelyn Ford's 2014 film *Nowhere to Call Home*, which focuses on a young Tibetan woman by the name of Zanta (like many Tibetans, Zanta only has one name) and her school-age son, Yang Qing. Originally from Tibet, Zanta had previously eloped with a man from a nearby village, with whom she had a son. After her husband died unexpectedly, however, her father-in-law tried to pressure her into marrying his other son, who at the time was in prison for murder. To avoid this remarriage, Zanta moved with her son Yang Qing to Beijing, where she managed to eke out a meager living peddling Tibetan beads, bracelets, and carved images of the Buddha. Zanta and her son encounter considerable discrimination in Beijing, as local biases against migrant workers are superimposed onto latent prejudices against Tibetans and other ethnic minorities. As a result, Zanta had difficulty even finding a landlord willing to rent them an apartment, and enrolling her son in school proved to be an even greater challenge.

At one point, an American radio journalist researching a story on migrant workers approached Zanta on the street, where Zanta was selling her wares. They chatted for a while, and before the journalist departed she gave Zanta her telephone number, urging her to stay in touch. Later, and seemingly out of the blue, Zanta reached out to the journalist again, asking her to adopt Yang Qing as her own son. Zanta explained that, ever since the day the two women first met, Zanta had been convinced that they must have known each other in a prior life, which is why she then resolved to hand over her only son to this woman she barely knew. In addition, it turned out that Zanta, who by that point had not yet managed to enroll Yang Qing in a local school, hoped that by finding him an American foster mother she would be able to provide him with more opportunities to pursue an education. The journalist was deeply moved by this gesture, and while she declined Zanta's request that she adopt Yang Qing, she nevertheless agreed to pay the tuition and other fees that would be necessary to get him enrolled in school, and furthermore she decided to film a documentary about the family's struggles The journalist is Jocelyn Ford, the former Beijing bureau chief for U.S. public radio's *Marketplace*, and the result of her collaboration with Zanta is the film *Nowhere to Call Home*.

Ford admits that the resulting project placed her in a rather complicated position. As a journalist, she had been trained to maintain a critical distance from the subjects on which she was reporting—though in this case she felt she was one of the few people capable of offering Zanta the assistance she desperately needed. As Ford recalls, "I was trying to be a journalist and stay out of it for a long time and not impose my views.... But at one point, I just could not stand it anymore, because nobody in Zanta's family was standing up for her" (Ford 2014). The result is a work in which the filmmaker attempts to draw attention to a set of structural inequalities in contemporary China, even as she was documenting her own attempts to address these same inequalities. The story she tells is one of intersecting issues of ethnicity, gender, and migration in contemporary China, though its specific configurations are inevitably shaped by her own involvement in the family's fate.

The question of Ford's involvement with Zanta's family becomes most acute when, near the end of the film, Zanta travels home with her son for the Lunar New Year holiday, whereupon her father-in-law seizes Yang Qing and refuses to return him to her. Driving this confrontation are competing visions of the future, with Zanta's father-in-law insisting that Yang Qing stay with him in Tibet to carry on the family line, while Zanta wants to provide her son with the education she hopes would allow him to pursue a greater range of opportunities beyond Tibet after he grows up. To help resolve the stand-off, Ford helps Zanta draw up a contract for the father-in-law to sign—whereby he would agree to permit Yang Qing to return to school in Beijing if Zanta, in turn, promised to bring Yang Qing back to visit his grandfather on a regular basis, thereby fulfilling his filial obligations.

The resulting scene of Zanta's father-in-law being given the contract is oddly poignant, as the older man—who had been presented as a tyrannical figure throughout much of the film—suddenly appears quite vulnerable when being asked to sign a document written in a language he cannot even read. This contract would embed the father-in-law within a legal structure with which he has little experience, just as the education Zanta is trying to provide for her son would embed him within a social system to which he has had only limited exposure. Yang Qing's mother and grandfather each have very different visions of the boy's future, and both rely on a set of extrapolations based on their knowledge and experience. In the documentary, we hear a lot from Zanta, Zanta's father-in-law, and even Ford herself regarding their respective visions of the boy's future. Curiously, however, the person from whom we hear the least is actually Yang Qing himself, who remains a cipher throughout much of the work. What are Yang Qing's interests and desires in this struggle over his future? Is it even possible to consider those interests and desires independent of the various familiar and societal forces that are simultaneously attempting to intervene in Yang Qing's fate in different ways?

Last Train Home

Released five years before *Nowhere to Call Home*, Li Xinfan's 2009 documentary *Guitu lieche* (Last train home) focuses on a teenage girl by the name of Zhang Qin and her younger brother, Yang, who are both being raised by their grandmother in a village in Sichuan Province while their parents work in a textile factory in Guangzhou. The film follows the family from 2006 to the summer of 2008, focusing on the period each year when the parents return home for the Lunar New Year. The result is a set of peculiar homecomings, as the parents visit children who increasingly view them as strangers and who chafe at the parents' attempts to make decisions about their future.

The discussions at these annual reunions invariably cycle back to the issue of the children's education. The parents are convinced that education is the best way for their children to secure a better future, but Qin has little interest in the education offered by her local schools. For her, the countryside is a place of social death, where her desires will always be subject to the authority of others—including her parents, her grandmother, and her teachers. Instead, she longs to relocate to the city and the modernity that it represents. The urban factory work that her parents regard as merely a means to an end is, for Qin, a desirable end in its own right.

Directed and produced by the Canada-based Chinese filmmaker Fan Lixin, *Last Train Home* is structurally aligned with the attempts by Qin's parents to provide her with a conventional education, though the film often appears to be more sympathetic to Qin's own distrust of those plans and her own desire instead to relocate to the metropolis. Qin is eventually allowed to accompany her parents to Guangzhou, where she too begins working in a factory. Although the factory environment is presented as being fairly confining, with workers forced to log long hours and live in cramped dormitory-style settings, Qin nevertheless appears to find the environment quite liberating. We see her chatting happily with her coworkers in their dorm room, going shopping for new clothes, and getting her hair curled at a salon. In fact, she is presented as enjoying a lifestyle not unlike that of the college student her parents had hoped she might become.

At the end of Qin's first year in the city, she and her parents find themselves trapped in the notorious 2008 blizzard that knocked out electricity and left the rail system in large portions of the country virtually paralyzed just as tens of millions of migrant workers were attempting to make their annual pilgrimage home for the New Year holiday. In one scene, we see Qin and her parents in the middle of a vast crowd of people waiting anxiously to see if there would be any train tickets available that day. The camera pans over a number of unidentified travelers before returning to the family, who are still waiting for tickets. Qin had evidently just been laughing at the absurdity of the situation, and now her mother is reprimanding her for her disrespect. Though we don't see Qin's initial reaction (she is no longer laughing by the time the camera pans back to her), this exchange nevertheless captures the stark contrast between the spirit of adventure with which Qin views her stay in Guangzhou, on one hand, and the sense of selfless obligation that has driven the parents' dedication to working in the factories for so many years, on the other. For Qin's parents, the many years of factory work in the south is merely a means to an end, while for Qin this same factory work (and all of the hardships it entails) is a desired end in its own right.

Tensions between Qin and her parents erupt most dramatically in a scene that takes place after the family finally makes it home in the winter of 2008. Qin and her father get into a heated argument over her future, in which Qin's father criticizes her for simply wasting money and not having a plan for what to do with her life. Qin responds angrily to these criticisms, whereupon her father orders her out of his house, and she curses back, "I'll walk right out of your fucking house."[5] The father immediately springs to his feet and begins slapping her, as Qin continues defiantly repeating the word "fuck" (*cao*) over and over again. The two of them struggle for several seconds, until Qin falls to the floor. Meanwhile, the filmmaker, evidently taken by surprise by this sudden outburst, struggles to keep the action in view, and in the resulting chaos the camera's boom microphone is inadvertently revealed in the upper portion of the screen—reminding the audience of the intrusive presence of the filmmaker in this intimate domestic scene. While there is no indication that the family is performing for the camera, the appearance of the boom microphone functions as an important reminder that, while the cinematic apparatus is largely invisible to the film's audience, it is nevertheless almost continually visible to the family themselves.

The presence of the filmmaker in this scene is made even more evident about a minute later, when a tearful Qin turns to the camera and exclaims, "You want to film the real me? This *is* the real me. What else do you want?" Spoken directly to the camera—and also, by implication, to the film's future audience—Qin's question breaks down the film's imaginary fourth wall and also underscores one of the central concerns that run through the work as a whole. In particular, the documentary revolves around the sacrifices that Qin's parents make in order to help provide their children with a better future, while at the same time highlighting the ambivalence with which Qin herself views this objective. For Qin, going off to work in the city is not so much a means to an end as it is an end in its own right, representing the possibility of escaping the dreary rural life within which she feels trapped. At issue, in other words, is precisely how one understands Qin's "real me"—the needs and desires of the young woman on whose behalf her parents, and even the filmmaker himself, are laboring. Is it even possible to conceive of Qin's "real" identity independent of the sociocultural networks within which she is embedded? Conversely, what role might the documentary apparatus have played in spurring Qin to articulate this defiant assertion of her "real" identity?

One of the final sequences in *Last Train Home* features Qin back in the city watching a television broadcast of the opening ceremony of the Beijing Olympic games with a number of other young people, who appear to be coworkers, while in the following scene we see her dancing in a nightclub. The significance of the film's nightclub scene is ambiguous, but given the enduring stereotype in contemporary China, associating migrant women and sex work, the scene suggests the possibility that Qin may be working as a consort. This nightclub scene, meanwhile, is juxtaposed with a sequence focusing on Qin's father, who at this point is in poor health and has been briefly hospitalized, and Qin's mother, who wonders tearfully how much longer they can keep up this sort of lifestyle. Their sense of quiet desperation stands in stark contrast to the optimism and enthusiasm evidenced by Qin, on whose behalf the parents are ostensibly working in the first place.

When the Bough Breaks

While *Nowhere to Call Home* and *Last Train Home* both revolve around parents attempting to support their children's educational prospects, Ji Dan's 2011 documentary *Wo chao* (When the bough breaks) instead depicts a migrant family where the parents have little interest in their children's education, and instead it is left up to the children to find alternative options when the migrant school where they are enrolled is scheduled to close.

The family lives in a makeshift hut on the outer fringes of Beijing, surrounded by mountains of trash that they and other migrants pick through, looking for items that may either be used or sold. The father is a cripple who lost part of his left leg in an accident and who suffers excruciating back pain for which he has little access to medical service in the city, given that he doesn't have a Beijing *hukou*. He initially comes across as a pathetic figure with limited physical mobility and even less social mobility, though as the film progresses he increasingly emerges as an alcoholic tyrant who continually berates his wife and children. He forces his family to perform humiliating rituals such as clipping his toenails, while rhapsodizing about the glories of the socialist past. The children appear to chafe under his sadism but nevertheless continue to respect his parental authority.

The family has four children, including an elder daughter who, we are told, previously left school to work but then disappeared, and the family is worried that she has probably been sold into the sex trade. The elder sister never appears in the film in person, but her absence haunts the family and gives added urgency to her younger siblings' attempts to negotiate their own futures. The remaining siblings include two sisters: Ling, who is comparatively quiet, and Xia, who aggressively attempts to step into the leadership role that has been vacated by the children's parents. In particular, one of Xia's primary concerns centers around the issue of education, given that the migrant school where the three younger siblings are enrolled on tuition fellowships is about to be demolished, leaving them with few viable alternatives. Interestingly, however, Xia's considerable energy and determination end up being directed toward securing educational possibilities not for herself and her sister, but rather for their younger brother, Gang, who has been offered the opportunity to attend an experimental private school in distant Hebei Province, for which he would need to pay RMB 13,000 in tuition. His family, however, has only RMB 9,000 in savings, and Xia is at her wits' end trying to find a way to make up the difference. We repeatedly see her literally pounding her head in frustration, as she tries to think of a way of earning thousands of extra RMB out of the scrap heap where her family lives. As Xia comes to assume the financial responsibilities that would normally fall under the purview of her parents, however, she increasingly models herself on her father, barking angrily at the siblings she is ostensibly trying to help. In this respect, she appears to replicate, at an unconscious level, the family's practical and emotional constraints, even as she is simultaneously trying to grant her brother an opportunity to escape those same constraints.

At one point, it is revealed that a certain Mr. Chen, who never appears in person in the film, has offered to provide several thousand yuan for the girls' tuition. The girls' mother is enthusiastic about this opportunity, and she encourages her daughters to take Mr. Chen up on his offer. What the mother does not realize (and what the film's viewers only learn through the work's editorial subtitles) is that Mr. Chen's offer is not mere charity, but rather is being offered in return for the girls' sexual services; he apparently suffers from an unspecified illness and has concluded that sleeping with virgins will provide a cure. The girls earnestly debate whether or not to accept the man's offer, though their focus is on how the money might be used not for their own education but rather for that of Gang.

Like *Nowhere to Call Home* and *Last Train Home*, *When the Bough Breaks* raises complicated questions about the ethics of observation and intervention. The filmmaker Ji Dan first met the family in 2004, when she was working on another project. Five years later, however, the children reached out to her again to tell her that their elder sister had disappeared and that the family was devastated by this development, and it was at this point that Ji Dan decided that the family might offer useful subject matter for a documentary. Unlike *Nowhere to Call Home*, however, the director never appears in person in the film, nor is there a moment, as in *Last Train Home*, when the cinematic apparatus becomes visible or when a character speaks back to the camera. Instead, the closest analogue in *When the Bough Breaks* occurs during a family argument over Gang's education, during which the father declares that he won't discuss the matter further, adding (with reference to the film camera) that there exists a record of his refusal to speak more on the topic. The irony is that the documentary's explicit reference to the cinematic apparatus occurs in the context of a discussion not of what is being recorded but of what is not.

Ji Dan worked closely with the children as she filmed the documentary, to the point of even moving into the teacher's dormitory at the migrant school where they were enrolled so that she could see them more often. The siblings frequently refer to Ji Dan on camera (affectionately calling her "Auntie"), though without specifying that she is filming them. In an interview conducted after the film was released, Ji Dan noted that she not only observed and recorded the family, but she also occasionally intervened more directly into their lives. In particular, while the RMB 4,000 (approximately USD 500) that Xia desperately attempts to raise in order to send Gang to his new school is clearly a vast sum for the migrant children, it is presumably a somewhat more accessible amount for a professional filmmaker like Ji Dan. In fact, she acknowledges in the interview that she did ultimately give the children several thousand RMB to support their education:

> You know, they're kids. They're just kids, so while I was filming, I really wanted to protect them. So, on the issue of money and specifically their need for tuition money, this was an extremely difficult thing to see happen. Their parents didn't help them. I was filming them and I really couldn't just sit back and watch them coming so close to a world of suffering and possible danger, given the lengths they were willing to go to secure the money. So, when I film a documentary, I feel that getting involved in this way is one of my strengths, something I can do to help. This is real life, you know. I don't have a lot of money, but these people had become my friends. They needed 3,000 RMB for the tuition. So, just as I would with any friend who really needed my help, I helped them and gave them 3,000. At the time I was filming, of course I had to consider how this would impact our relationship and the film, but I thought that to not help them would be a sin. It would be a sin to watch these children endure such suffering. It was about more than making a great film. (Rudolph 2012)[6]

This revealing statement, however, leaves a number of questions unanswered. Ji Dan states that it would be a "sin" to watch the children endure the suffering of struggling to raise the money they would need, though the documentary itself functions to document this suffering and present it to the film's audience. When precisely did Ji Dan decide to give the children the money? Did she deliberately wait until after she had enough footage of the children struggling to come up with a way to raise the impossible sum, or did she have them perform this desperation even knowing that Ji Dan would help them out? And if she does indeed deem it a "sin" to watch the children suffering in this way, then what are the ethics of recording this struggle and putting it on display for a mass audience?

Furthermore, Ji Dan's decision to give the children the money appears to accept and validate Xia's assessment of her family's needs, specifically the girl's decision to prioritize her younger brother's education over her own and that of her sister. But surely this tacit reaffirmation of a traditional androcentric attitude toward education is not the message that Ji Dan—herself one of contemporary China's leading female documentary filmmakers—is trying to convey with this film? Also, why does Ji Dan not make her intervention into the family's fate explicit within the documentary itself, the way that Jocelyn Ford does in *Nowhere to Call Home*? What might the work have looked like had she done so?

Conclusion

As it turns out, the Dandelion School in Beijing, where I have been working, did not close as a result of its inability to bring its buildings up to code. Instead, around the same time that the school was confronted with the challenge of renovating its substandard buildings, it also learned that Beijing's Nanyuan Airport was scheduled to close as soon as the new Beijing Daxing International Airport opens in 2018, whereupon the region around the old airport would be rezoned for residential housing. The Dandelion School, accordingly, made plans to move out of the neighborhood altogether and to relocate into a new building elsewhere in Beijing. As this volume goes to press (in 2016), the school is currently in its final year in its current location. What the future holds, no one can be certain.

The ways in which the fate of the Dandelion School has been buffeted by a set of overlapping sociopolitical tectonic shifts—ranging from the sociopolitical aftershocks set in motion by the 2008 Sichuan earthquake to Beijing's growing infrastructural requirements—provide an apt metaphor for the precarious fate of the tens of millions of children of China's migrant workers, as they struggle to find a place for themselves within this same rapidly shifting landscape. As members of China's next generation, these children represent China's future, even as their individual fates frequently depend on a set of developmental trajectories largely beyond their immediate control.

Each of these three documentaries attempts to document the challenges encountered by these migrant children, specifically in relation to the challenge of pursuing a formal education. These works are filmic explorations in migrant marginality, played out in dramas of family trauma, dysfunction, and discrimination against the rural migrant turned urban, who will always be treated as the outsider. They promote imagined futures predicated on the promise of China's education system, while also questioning what the ultimate value of that education might be.

Notes

1 The government's figure was actually slightly higher than the one Ai Weiwei and his research team had come up with by that point (a figure they were certain was a significant underestimate). Unlike Ai Weiwei, however, the government did not release individual names, so it was impossible to know whom precisely they were counting. For a discussion of Ai Weiwei's Sichuan earthquake activism, see Grube (2009). For a discussion of Ai Weiwei's politics in relation to his art, see Sorace (2014).

2 A review of the literature on migrant schools in Beijing and other urban centers in China is beyond the scope of the present discussion. Human Rights in China (2002) produced one of the earliest reports, which highlighted periodic school closures, practices of discrimination, and the general poor quality of the earliest migrant schools in Beijing. In the buildup to the Beijing Olympics, many schools were forcibly shut down and entire "migrant" neighborhoods were razed for the development of infrastructure projects such as roads, hotels, and high-rise apartments. A series of school closures came again in the summer of 2011, and again in 2014, as many mainstream newspapers reported. Since then, we have seen the publication of a number of book-length studies about migrant education, with much of the research focusing on the post-2008 situation. See Wu, Zhang, and Webster (2014); Ming (2014); Pong (2015); and Li (2015).

3 My time at Dandelion was enabled by the hospitality and collaboration of Zheng Hong, the principal of the school, and the school staff and teachers, many of whom are migrants themselves, and by the Duke Engage project, which allows me to send from six to ten Duke students to volunteer at the school each summer. The work at Dandelion has been part of a larger research project aimed to understand migrant education in relation to questions of urban development, labor, corporate and civic responsibility projects, and media representation. See Litzinger (n.d.).

4 To date, there have been few studies of representations of migrant children and the politics of schooling in documentary and other film genres and in the media more generally in China. Scholars have tended to focus more centrally on the deployment of images of migrant workers in film, photography, and art. See Sun's (2014)

groundbreaking study, *Subaltern China*. Other important studies include Kochan (2009) and Parke (2015).

5 Here and elsewhere in this chapter, the English translation of the films' dialogue may differ slightly from that which appears in the works' official subtitles.

6 Curiously, however, even though in the film itself the children repeatedly emphasize that they need RMB 4,000 to cover Gang's remaining tuition, here Ji Dan only mentions giving them RMB 3,000.

10

"I AM GREAT LEAP LIU!"

Circuits of Labor, Information, and Identity

in Contemporary China

Carlos Rojas

Yang Zhi knew that *Duanduan* must be a fake name. But if a name is used and mutually accepted, it thereby becomes real. Is it important whether a term of address is correct or not?
—Liu Zhenyun, *I Am Liu Yuejin*

"You don't *look* like Liu Yuejin!"[1]

A burly man from Gansu Province peers down curiously at his hapless victim, whom he has just caught in bed with a putative prostitute. The photo ID the Gansu man is holding carries the name *Liu Yuejin*, though the image in the photo bears little resemblance to the now-naked man in whose waist pack the ID was found. As a result, the ID functions here as a token not of identification but rather of *mis*identification—in that the subject's identity is located precisely in the mismatch between the photo and its ostensible referent.

It turns out that both the ID and the waist pack originally belonged to a migrant worker from Henan Province, who works as a cook for a Beijing construction site. The waist pack was stolen from the *real* Liu Yuejin by another migrant from Shanxi Province by the name of Yang Zhi, who himself almost immediately falls victim to a scam perpetrated by a group of migrants from Gansu. The latter scam begins when a young woman pretending to be a streetwalker lures Yang Zhi back to a one-room apartment at the end of a filthy alley just beyond Beijing's Fifth Ring Road, but soon after they get undressed and go to bed they are interrupted by the woman's three male partners, who suddenly barge into the room and confiscate all of Yang Zhi's possessions—including the waist pack he had pilfered from Liu Yuejin just hours earlier. The picture ID found in the waist pack, accordingly, comes to function not as an indicator of who Yang Zhi *is*, but rather as a counterindicator of who he *is not*.

This sequence, from the opening chapter of the 2007 novel *Wo jiao Liu Yuejin* (I am Liu Yuejin) by Beijing-based author Liu Zhenyun, is set against a backdrop of a shifting set of approaches to demographics and identity in contemporary China. Under the household registry system that was implemented in 1958, all Chinese citizens were assigned a residency permit, or *hukou*, that entitled them to social benefits such as access to public education, health care, jobs, pensions, and even grain rations. The system was designed to limit unauthorized movement of citizens, particularly from rural areas into the cities. During the period of rapid economic growth and political liberalization that began with Deng Xiaoping's 1978 Reform and Opening Up Campaign, however, a burgeoning wealth gap between rural and urban areas helped contribute to a socioeconomic environment in which many rural residents concluded that the potential advantages of coming to the city to look for work outweighed the loss of *hukou*-related benefits they would incur as a consequence of leaving home. The result has been the emergence of a vast floating population of around a hundred and forty million migrant workers. In Beijing alone, there are currently more than seven million migrant workers, or more than a third of the city's total population, and it is no coincidence that all of the characters introduced in this introductory sequence of Liu Zhenyun's novel are migrant workers who have come to Beijing to seek their fortunes.

China's migrant workers occupy a curious interstitial position in contemporary society. Even as the nation relies on migrants' cheap labor to fuel its continued economic growth, the state nevertheless relegates those same migrants to a sort of legal twilight zone in which they are tacitly encouraged to come to the cities to work, but are simultaneously prevented from claiming many of the benefits to which the city's legal residents are entitled. Although the weakened *hukou* system is no longer able to stanch a tidal flow of internal migration, the system does nevertheless continue to mark these migrants *as migrants*—condemning the vast majority to the status of permanent outsiders within the urban areas where they work and reside. In this respect, the primary function of the *hukou* system is increasingly to place migrant workers within a space of strategic misrecognition, much like the situation in which Yang Zhi finds himself in his encounter with the Gansu con men.

While it is certainly true that China's migrant workers occupy a very precarious position within this new economy, the recent weakening of the *hukou* system has nevertheless helped create an environment in which rural Chinese theoretically may come to have greater control over their own fate. Rather than remaining strictly defined by the biographical and residential information contained in their *hukou* files, these rural residents increasingly choose to pursue potentially more lucrative opportunities in the cities, where some of them may establish new homes, new families, and even new identities. In Liu's novel, this possibility of self-reinvention is concisely captured in the scene when Yang Zhi, upon first meeting the (fake) streetwalker, politely asks her name. The woman replies that her surname is Zhang, then adds "but you can just call me Duanduan." The narrator then observes, "Yang Zhi knew that *Duanduan* must be a fake name. But if a name is used and mutually accepted, it thereby becomes real. Is it really important whether a term of address is correct or not?"

What is being affirmed in this scene is a precise inverse of the Confucian advocacy of a rectification of names. The *Lunyu* (The analects) famously posits that names must be correct in order for language to be in accordance with reality, which in turn is regarded as a necessary prerequisite for having effective government and a harmonious society.[2] Contrary to this Confucian insistence that language and names be reconciled with an idealized social reality, however, Liu Zhenyun's narrator suggests that it is instead reality that will inevitably follow the lead of language. Social reality, under this latter view, is the product of a collaborative use of language, or what Jürgen Habermas calls "communicative action." Within the resulting fluid system of denotation and reference, individuals may encounter a reaffirmation of their accepted social identity (as is the case with many social elites), may find themselves occupying a space of strategic misrecognition (as does Yang Zhi, in his encounter with the Gansu con men), or may even attempt to claim a new identity altogether (as does Zhang Duanduan, in her encounter with Yang Zhi).

In the following discussion, I use Liu Zhenyun's novel to reflect on these shifting understandings of society and identity in contemporary China. I argue that the novel's migrant workers shuttle back and forth not only between different geographical locations, but also between different sociopolitical conceptions of individual identity. Drawing on psychoanalytic theories of trauma, economic understandings of debt, and philosophical models of denotation and reference, I contend that contemporary Chinese society is shaped by the mutual imbrication of two distinct models of social organization and information management implicit in the traditional Maoist *hukou* model and the contemporary neoliberal economy, respectively. The novel's migrant workers illustrate both the far-reaching ramifications of each of these socioeconomic models, as well as their respective limitations.

Traumatic Origins

I Am Liu Yuejin opens in medias res, with Yang Zhi having just finished lunch in a small Beijing restaurant owned and run by a certain Old Gan, who, like Yang Zhi, is a migrant worker from Shanxi. The narrative begins with Old Gan having just come over to bring Yang Zhi the check, whereupon he asks whether Yang Zhi has had any luck tracking down his, Old Gan's, leather jacket. The jacket had been stolen from the restaurant a few weeks earlier, and Old Gan had offered to pay Yang Zhi a twenty-yuan reward if he could use his connections to Beijing's underworld to help recover it. The jacket, we are told, was not particularly valuable in its own right—and, in fact, it wasn't even made from real leather—but Old Gan desperately wants it back because in one of the pockets there was a small notebook containing a record of all of the tabs his customers had run up. The problem is not that Old Gan doesn't know who owes him what (he actually knows the amounts of all the debts like the palm of his own hand), but rather that he is concerned that without this physical record of the debts his customers will refuse to pay what they owe. When Old Gan brings up the stolen jacket, however, Yang Zhi spits on the ground in disgust and complains that Old Gan is requesting his help while at the same time asking him to settle his bill. Yang Zhi nevertheless pays for his meal and leaves the restaurant, and it's at this point that Zhang Duanduan follows him out into the street and proceeds to offer her services.

While Old Gan's leather jacket and account book are never mentioned again in the novel, this opening vignette introduces the work's more general concerns with issues of trauma, debt, loss, and mediated returns. Following his encounter with the Gansu con men, for instance, Yang Zhi becomes distraught about his loss not of the waist pack (which, after all, had only been in his possession for a few hours before it was stolen again), but rather of his virility. That is to say, after being interrupted in flagrante delicto by the con men, Yang Zhi discovers that he has been rendered impotent, and he concludes that his only hope of recuperating his libido lies in tracking down Zhang Duanduan. Yang Zhi's determination to find Duanduan may therefore be seen as a version of what Freud calls a "traumatic neurosis," or a compulsion to reenact the circumstances of an earlier traumatic event. According to Freud, this process of compulsive reenactment functions to make the original trauma intelligible, and by this logic Yang Zhi's quest for Duanduan may be seen as an attempt to reconstitute the significance of the original encounter (Freud 1920; Freud 1937–1939).

Even as Yang Zhi is trying to track down Zhang Duanduan, Liu Yuejin is himself urgently searching for Yang Zhi in order to regain his stolen waist pack, the personal significance of which is similarly rooted in an earlier moment of sexual humiliation. It turns out that, in addition to Liu Yuejin's ID card, address book, a handful of receipts, and about 4,000 yuan in cash, the stolen waist pack also contained his divorce certificate and a promissory note for 60,000 yuan. The significance of the promissory note lies in a sequence of events that had unfolded six years earlier. At the time, Liu Yuejin had just accepted a new job working as a cook in the Henan county seat but his wife was still living back in the countryside, and Liu therefore asked his new boss, Li Gengsheng, to arrange for Liu's wife to join him in town. Although Li Gengsheng agreed to help bring Liu's wife over, he also began sleeping with her soon after she arrived. When Liu Yuejin subsequently discovered the affair, he demanded not only that he receive an immediate divorce but also that Li Gengsheng compensate him for his ensuing "psychological distress." Li agreed, but only on the condition that the money be put in escrow for six years and that the payment ultimately be contingent on Liu's remaining on "good behavior"—which is to say that Liu not interfere with Li's relationship with Liu's (ex-)wife following the divorce. For Liu, therefore, the IOU functions both as a reminder of his earlier betrayal as well as an anticipation of the new life he hopes to establish once he receives the settlement. By keeping Liu tethered to the memory of the original separation, the IOU invites, or even requires, a version of the sort of compulsive repetition that, Freud argues, retrospectively writes the earlier trauma into existence. As Cathy Caruth summarizes, a trauma is an attempt to convey "a reality or truth that is not otherwise available" and which, "in its delayed appearance and its belated address, cannot be linked only to what is known, but also to what remains unknown in our very actions and our language" (2010: 4–5).

In the figure of the IOU, the legacy of Liu Yuejin's divorce is explicitly translated into a monetary debt. Like a Freudian trauma, a debt occupies a space of anticipatory potentiality, wherein the significance of the initial economic encounter is not fully realized until a future point in time. The specific terms of Li Gengsheng's contract, moreover, place Liu Yuejin in an odd temporal interregnum—expressly forbidden from reengaging with his former life prior to the divorce, but at the same time forced into a holding pattern until he can receive the funds that might permit him to establish a new life for herself. Even after Liu relocates from the county seat to Beijing, accordingly, the IOU continues to mark a set of complex emotional and financial ties linking Liu to his ex-wife and her new family.

The IOU's double significance as a symbol of Liu Yuejin's past *and* his future is further complicated when, approximately halfway through the novel, Liu is unexpectedly visited by his son, Liu Pengju, who by this point is already a teenager. In negotiating the divorce agreement, Liu Yuejin had not only demanded sole custody over Pengju, he had even refused, on principle, to let his ex-wife contribute to Pengju's tuition. When Liu subsequently moved to Beijing, however, he left Pengju enrolled in a private school back in the county seat, and consequently he is startled when Pengju suddenly shows up at his door to announce that he has quit school and wishes to relocate to Beijing, where Pengju and his girlfriend plan to open a pedicure parlor. Having heard Liu Yuejin repeatedly brag about being in possession of more than 60,000 yuan, Pengju asks Liu to give him money for his new business—without realizing the specific source or significance of the money in question, or the fact that Liu Yuejin doesn't yet have access to the actual funds.

Like the IOU, Liu Yuejin's son, Pengju, is a material embodiment of both the legacy of Liu's former marriage as well as his hopes for the future. These contradictory associations may be observed in Liu Yuejin's earlier insistence on being solely responsible for Pengju's tuition following the divorce, even as he was simultaneously demanding that Li Gengsheng pay him a large sum in compensation for the distress he suffered as a result of the adulterous affair that made the divorce necessary in the first place. Each of these seemingly antithetical requests, moreover, had both symbolic and practical implications. Liu's demand that Li Gengsheng pay him the 60,000 yuan, for instance, had the symbolic effect of punishing Li Gengsheng for his actions and also offered the practical advantage of providing Liu with a solid financial cushion with which he might establish a new career and a new family. Conversely, Liu's insistence on being solely responsible for Pengju's tuition not only functioned to reaffirm his identity as a responsible father, but also was a calculated investment in his son's (and, by extension, Liu's own) future. By paying for Pengju's education, in other words, Liu Yuejin was not only increasing Pengju's chances of securing a more lucrative career when he grows up, but also hopefully buttressing the likelihood that Pengju would feel a filial obligation to share that future wealth with Liu himself.

The choices that Liu Yuejin faces with respect to his relationship to his son and ex-wife mirror a set of similar challenges being negotiated by countless other migrant workers throughout China. While some migrants do bring their entire families with them when they relocate to the city, many others are forced to leave their children behind precisely in order to be able to better provide for their tuition and upbringing, or alternatively they send the children themselves into the city to find work and help support the family. The irony is that, in attempting to provide more reliable material benefits for their families, these rural parents are frequently forced to distance themselves from their own children, which may well contribute to feelings of estrangement and alienation that can eventually tear the families apart.

Two recent documentaries on Chinese migrant laborers explore these issues from different perspectives. As Ralph Litzinger discusses in his chapter in this volume, in the 2009 film *Guitu lieche* (Last train home, dir. Lixin Fan), we witness a growing emotional rift between a teenage girl, living in rural Sichuan, and her parents, whom she sees only once a year when they return from Guangzhou, where for years they have been working in a factory. The 2005 film *China Blue* (dir. Micha Peled), meanwhile, examines a similar phenomenon but focuses instead on a group of teenage girls who have been sent by their families to work in factories in the city in order to help supplement their families' income. In both instances, the goals and dreams of each generation diverge markedly from one another—a process that is frequently exacerbated precisely by the family members' attempts to make those same dreams a reality. In particular, a recurrent issue explored in each of these works is whether these goals are best conceived narrowly, as relating primarily to oneself, or more broadly, as embracing the entire family unit within which one is positioned.

In *I Am Liu Yuejin*, meanwhile, we find a version of this sort of fractured family structure in Liu Yuejin's complicated relationship with his son, Pengju. The day after Pengju and his girlfriend arrive in Beijing, they abruptly leave again, returning to the county seat where Liu Yuejin's ex-wife lives with her new family. After failing to secure the desired start-up capital from Liu Yuejin, Pengju subsequently discovers that the money in question is actually still in the hands of his stepfather, Li Gengsheng. Pengju therefore asks Li himself for the money, and when Li refuses Pengju proceeds to take his own half-brother hostage—this being the newborn son born to Pengju's mother and her new husband, Li Gengsheng. Using his baby brother as blackmail, Pengju demands that Li Gengsheng give *him* the 60,000 yuan that Li technically owes Liu Yuejin. In treating his infant half-brother as collateral with which to demand from his stepfather the monetary settlement that is technically owed to his biological father (but which Pengju has concluded is rightfully his), Pengju is tacitly underscoring the overlapping rifts that divide this fractured family and their shattered dreams. In holding his infant brother as collateral for securing the monetary sum with which he hopes to establish his future career, Pengju is figuratively working through his own experience of having been simultaneously claimed and abandoned by his father, Liu Yuejin, at the time of his parents' divorce. In using this negotiation with his stepfather and infant half-brother as a way of revisiting, in a mediated fashion, his conflicted relationship with his own biological father, meanwhile, Pengju may be seen as retrospectively granting meaning to a traumatic legacy that previously had not even been intelligible as such. In particular, by using his half-brother to blackmail his own stepfather in order to secure the money intended for his biological father, Pengju is emphasizing the degree to which Liu Yuejin's original insistence on paying for Pengju's education itself could be seen as a form of blackmail, underscoring the son's filial obligation to reciprocate in kind in the future.

Archived Debts

As Liu Yuejin is attempting to locate his stolen waist pack (and, more important, the promissory note it contained), he happens upon an expensive purse that Yang Zhi had recently stolen from the wife of a wealthy businessman but then immediately abandoned. The purse contains a cosmetics set, some bank cards, about five hundred yuan in cash, together with a computer flash drive, and it subsequently turns out that the files stored on the flash drive include incriminating photographs, audio recordings, and videos of many of the businessman's professional associates. The businessman had compiled these files for use in potentially blackmailing his associates, and his wife had secretly made an extra copy for her own protection. After the wife's copy is lost when her purse was stolen, the businessman becomes desperate to recover it, even offering a private detective 200,000 yuan to do so as expeditiously as possible. Eventually, a group of gangsters also join the search, as do some of the businessman's own associates. Liu Yuejin himself, meanwhile, doesn't realize the flash drive's significance until after it has already been stolen from him by

Pengju and his girlfriend, but upon discovering that everyone searching for the drive is convinced that he still has it, Liu Yuejin decides to play along in an attempt to maximize his own leverage (and, in the process, hopefully increase his chances of recovering the stolen IOU, which remains his primary objective). The result is a complex shell game, in which multiple parties comb the city in search of a pair of items that are themselves in a continual state of circulation.

The IOU that Liu Yuejin loses and the flash drive that he subsequently finds occupy almost precisely parallel positions within the novel. Both items are contained in stolen bags (and the novel frequently refers to both the waist pack and the purse as simply *bao*, or "bag"), and in each case they are seemingly inconsequential items whose value is subsequently revealed to be a speculative product of collective obligation and desire. In narratological terms, both items may be seen as examples of what Hitchcock would call a MacGuffin—which is to say, a narrative element that is presented as an object of desire, but whose primary function is simply to help drive the plot of a work forward. Beyond their role in the actual narrative, however, the IOU and the flash drive each have distinct implications for the shifting socioeconomic structure of contemporary Chinese society.

First, like Old Gan's stolen account ledger with which the novel opens, both the IOU and the flash drive are records of debt and obligation. As a product of the affair that destroyed Liu's marriage, for instance, the IOU is literally a contract for future compensation, which at the same time permits Li Gengsheng to control Liu Yuejin's behavior. The flash drive, meanwhile, contains personal information that may be used to blackmail others, and consequently it functions as an anticipatory debt that is explicitly designed to shape people's behavior. In this way, the drive becomes a potent symbol of the degree to which capitalism itself is driven by a parallel quest for information and an attempt to hedge against the dangers of misinformation—a symptom, in other words, of the speculative economy within which the drive itself is positioned.

Second, both the IOU and the flash drive are also literally repositories of information, though in each case the information in question is already available elsewhere (the IOU is a material record of a debt that is already known to both parties, while the flash drive contains merely a *copy* of a digital archive). There is, however, a key difference between the two items, in that the IOU is a signed contract whose significance lies precisely in its singularity, while the flash drive contains digital files that in theory can be copied and recopied ad infinitum. In this respect, the novel's juxtaposition of the IOU and the flash drive underscores an important shift in modern China between two distinct approaches to the management of personal information. On one hand, the IOU resembles the individual dossiers used under the old *hukou* system, wherein people's physical files were typically held by their work unit and their ability to relocate to other regions was constrained by their need to retain access to these same files. On the other hand, the flash drive is a product of a contemporary information system that places a significantly greater emphasis on vast archives of digital data. These digital archives may contain significantly more personal information than the old *hukou* dossiers, and unlike the traditional dossiers the digital data are not stored in any single location and instead may circulate widely in various forms and media.

In recognition of the growing importance of these sorts of digital data, China in the 1990s began to implement a set of major telecommunications and information infrastructure initiatives known as the Golden Projects, each of which consists of a set of networked databases covering areas such as customs, tax collection, intelligence, agriculture, and health. Collectively, these projects constitute an ambitious effort on the part of the government to systematize its handling of a variety of different digital archives, together with the populations to which they correspond. The most notorious of these Golden Projects is the Golden Shield Project—better known by its more familiar nickname, the Great Firewall of China. Initiated in 1998, just four years after China first connected to the Internet, the Golden Shield Project monitors information circulating over the Internet, while also selectively blocking specific transmissions and even entire websites and web domains. While this monitoring system is quite extensive, and has received considerable attention both in China and abroad, it could nevertheless be argued that one of its primary functions is not direct oversight but rather encouraging Chinese Internet users (both individuals as well as entire corporations) to engage in a process of self-censorship—preemptively responding and adapting to what they anticipate the government would expect of them.[3] Some observers have characterized the resulting system as being a sort of "digital panopticon," wherein users are never certain precisely when they are or are not being observed, and consequently they are encouraged to internalize the government's disciplinary gaze and modify their behavior in response. On the other hand, the government's oversight over the Internet has also helped inspire a complex ecology in which many Chinese netizens preemptively resort to euphemisms and other codes in order to circumvent the censors. In addition, in a bit of a double twist, it has even been argued that the government may be tacitly encouraging some forms of online dissent in order to help prevent those same critiques from being manifested in a potentially more destabilizing form, such as a mass protest.[4]

In its focus on the flash drive as a repository of personal information, *I Am Liu Yuejin* also builds on a thematics that Liu Zhenyun had previously explored in his 2003 novel *Shouji* (Cell phone). This earlier work describes how a Beijing talk-show host's life begins to unravel after his wife inadvertently discovers evidence of an extramarital affair from text messages on his cell phone. Like *I Am Liu Yuejin*, *Cell Phone* was released almost simultaneously with its cinematic adaptation (for which Liu Zhenyun also wrote the screenplay). Directed by Feng Xiaogang, one of contemporary China's most consistently successful directors, the cinematic version of *Cell Phone* was the nation's highest-grossing film in 2003, and it helped catalyze an animated debate in China over the effects of cell phones and other communication technologies. After watching the film, countless Chinese cell-phone users claimed to have realized for the first time that their phones contained potentially damaging personal information, and they hastily changed their phone-use habits accordingly. Given that the work ostensibly presents a critique of the detrimental impact of cell phones on social relations, it is ironic that one of the biggest corporate sponsors of the film was the cell phone manufacturer Motorola. Furthermore, the resulting publicity from the sponsorship played an important role in helping revive Motorola's slumping fortunes (Wang 2010). To the extent that *I Am Liu Yuejin* sets up an explicit parallel between the overlapping quests for the IOU and the flash drive, with the former symbolizing the traditional *hukou* and its relevance to the lives of contemporary China's underclass of migrant workers and the latter symbolizing modern information technologies and their relevance to the urban elite, such as the businessman trying to regain the flash drive, Liu Zhenyun's *Cell Phone* is a useful reminder that, at a time when the number of cell-phone accounts in China has already exceeded a billion[5] and the number of Internet users is in excess of half a billion,[6] the sorts of digital archives symbolized by the missing flash drive are ones that are directly relevant to broad swaths of the Chinese population. Although *I Am Liu Yuejin* might appear to be structured as a contrast between two very different social classes (poor migrant workers and wealthy businesspeople), accordingly, it in fact focuses on two systems of information and social organization (*hukou* and virtual archives) whose relevance extends to much of China.

The contrast between the two approaches to information and identity exemplified by Liu Yuejin's IOU and flash drive may also be examined through the lens of a debate in analytical philosophy concerning how proper names (and natural kind terms, such as gold) denote their referents. Around the turn of the twentieth century, the philosophers Gottlob Frege and Bertrand Russell proposed different versions of what is often characterized as a descriptive model of reference, wherein proper names are viewed as a sort of conceptual shorthand for a set of descriptive attributes corresponding to a certain referent (see Frege 1952; Russell 1956). A proper name, under this model, does not denote its referent directly but rather specifies a set of descriptive attributes associated with the corresponding referent. An alternative model subsequently popularized by Saul Kripke (1980), meanwhile, holds instead that names are "causally" linked to their referents by an initial "baptismal" act—after which the name and its referent remain tethered to one another, even if one's understanding of the name and/or referent happens to radically transform in the interim. Under this latter model, a proper name functions as what Kripke calls a "rigid designator," in that it continues to designate the same referent in all possible worlds in which that referent exists. If, for instance, it were to be proved that all of the works currently attributed to Shakespeare had actually been written by Sir Francis Bacon, the descriptivist model would predict that the name *William Shakespeare* would then come to refer to the individual formerly known as *Francis Bacon* (since he would be the person who would most closely match the attributes we currently associate with Shakespeare), while the causal model would predict that the name would instead continue to refer to the same individual (despite the fact that he is no longer who we originally thought he was).

More recently, in *The Sublime Object of Ideology*, Slavoj Žižek examines the relationship between descriptivist and causal (or what he calls antidescriptivist) theories of reference, and he concludes that they both "*miss the same crucial point*—the radical contingency of naming" (1989: 92; emphasis in the original). More specifically, he argues that what ultimately "guarantee[s] the identity of an object in all counterfactual situations—through a change of all of its descriptive features—is *the retroactive effect of naming itself*: it is the name itself, the signifier, that supports the identity of the object" (1989: 94–95; emphasis in the original). Like a Freudian understanding of trauma, in other words, the epistemological stability of a name's referent is established only retrospectively, from the perspective of the contemporary moment of utterance.

It is important to note, meanwhile, that what is ultimately at stake in these philosophy of language debates is not an intrinsic property of names themselves, but rather how people actually *use* these names. It is entirely possible that intuitions about naming and reference may shift over time and may in turn be shaped by the specific social and institutional structures within which they are deployed. In fact, there is evidence that cultural differences may lead people to have different intuitions about how names work, even among speakers of the same language.[7] Given the importance of these questions of how language is actually used, accordingly, it is arguably more useful to consider the ways in which different assumptions about naming and reference may inform real-world issues such as how people think about the relationship between individual identity and officially generated documents like ID cards and *hukou*s (as opposed to counterfactual scenarios such as the Shakespeare/Bacon example). From this perspective, people holding "causal" intuitions about naming are not only likely to view a proper name as continuing to refer to the original referent even when they are confronted with new information regarding the identity of the referent (as Kripke's model would predict), they may conversely also try to ensure that the referent will continue to conform to the information originally associated with the name to begin with. A "causal" approach to reference, in other words, may well have implications not only for how names are linked to their referents, but also for how referents may remain constrained by the information and assumptions embedded in the names themselves. Conversely, a "descriptivist" approach to reference may well encourage a more flexible approach to identity, by underplaying these same figuratively constraining links between names and their referents. The implication is that while a causal approach may mean that names will remain tied to their referents in a wide array of possible universes in which the understanding of those referents might undergo radical change, in more practical terms a causal approach may well *restrict* the ability of actual individuals to stray far from the specific attributes with which they are originally associated, and vice versa.

To the extent that the IOU and the flash drive, in Liu Zhenyun's novel, are metonyms of a broader contrast between two different ways of approaching the relationship between information and identity in contemporary China, we may borrow the terms of the preceding philosophy of language debates to help reframe the contrast between two corresponding approaches to society and governance. The traditional *hukou* system, on one hand, relied on an understanding of social identity that could be seen as basically causal (or antidescriptivist), insofar as individuals remained defined by the identity to which they were assigned by their *hukou*. The weakening of the *hukou* system in contemporary China, on the other hand, has yielded a more fluid situation analogous to what one might expect based on a descriptive theory of reference, insofar as people's social identity is increasingly informed not by the information specified in their *hukou* but rather by the shifting realities of where they actually end up living and working. Under this latter system, identity is determined not so much by a set of preestablished attributes, but by a complex array of data points that may be continuously revised and updated.

At the same time, however, even with the weakening of the *hukou* system, there remain significant constraints on the ability of migrant workers to create entirely new identities for themselves. In practical terms, even as migrants are able to modify certain aspects of their identity as specified in their *hukou* (particularly with respect to the region where they are assigned to live and work), they are less able to alter other attributes such as their gender, ethnicity, or proper name. Moreover, while it is certainly possible for many migrants to earn more money in the city than they would have had they remained in the countryside, it nevertheless remains difficult for them to significantly change their class position. Indeed, social mobility (defined as the ability of individuals to move from one socioeconomic tier to another) remains relatively low in China compared to other countries, due in part to China's vast rural population and the residual effects of the *hukou* system. If one looks at the situation intergenerationally, however, a different picture emerges, and studies indicate that, in contemporary China, social *fluidity* (defined as the ability of individuals to occupy a different socioeconomic tier than the one their parents belong to) is actually comparatively high (Wu and Treiman 2007). That is to say, while it remains relatively difficult for most rural residents to dramatically improve their own socioeconomic position, they are more likely to be able to help provide better opportunities for their children. The result, for many rural residents, is a disjunction between their ability to improve their own lot and their ability to improve that of their children.

These sorts of conflicting generational imperatives may be observed in Liu Yuejin's relationship with his son, Pengju. After Pengju steals the purse containing the flash drive, a group of gangsters proceed to kidnap him and his girlfriend, torturing them in an attempt to force them to reveal the drive's location. Once the gangsters determine that Pengju no longer has the drive, however, they instead use him as a hostage to force Liu Yuejin to find the drive and hand it over to them. Eventually, Liu Yuejin manages to use a decoy—a blank flash drive that looks exactly like the one that was lost—to trick the gangsters into releasing Pengju and the girlfriend. When Pengju subsequently kidnaps his infant brother in order to pressure his stepfather into giving *him* the 60,000 yuan that the stepfather technically owes to Liu Yuejin, Pengju is essentially taking his own experience of having been kidnapped and held hostage and subjecting his own kid brother to the same experience. In this way, Pengju implicitly embraces his own commoditized status, using it strategically to obtain the divorce settlement funds technically owed to his father, and with which he hopes to start his own business. Just as Liu Yuejin successively used a blank flash drive as a decoy in order to secure his son's freedom, Pengju similarly uses his infant brother as a stand-in for the written contract that his stepfather had been demanding in return for the 60,000 yuan settlement.

Deictics and Dialectics

The eponymous protagonist of Liu Zhenyun's *I Am Liu Yuejin* has an unusual given name, which means "leap forward." Given that in the Maoist era it was common for children to be given patriotic names inspired by the political discourse of the day, it is likely that the fictional Liu Yuejin would have been named by his parents after the Great Leap Forward (1958–1961), a political campaign known in Chinese as *da yuejin*.[8] Originally conceived as part of China's Second Five-Year Plan, the Great Leap Forward sought to jump-start the nation's transition from an agrarian to a communist economy by promoting a process of rapid industrialization, during which communities were encouraged to form communal collectives, thereby severing peasants' ties to their family-owned plots of land. The campaign, however, ultimately proved catastrophic, as pressure to meet unrealistic production quotas encouraged widespread fraud and led to enormous economic and environmental devastation, the result of which was a vast famine that claimed the lives of tens of millions of people. Indeed, in tacit recognition of the campaign's adverse consequences, the Chinese government, in a sort of retrospective "rectification of names," now refers to the corresponding period as the "three years of natural disaster" (*san nian ziran zainan*). Even though the political campaign after which Liu Yuejin was presumably named is one whose significance has been radically transformed by subsequent historical developments, Liu himself has nevertheless retained his original name—and with it both the optimistic connotations the movement originally carried as well as the more sober associations it subsequently came to acquire in historical hindsight.

While at no point does Liu Zhenyun's novel explicitly refer to the Great Leap Forward itself, the work does reflect indirectly on the historical campaign through its focus on the contemporary period, which has arguably managed to realize many of the original campaign's stated objectives. Whereas Mao

Zedong famously predicted at the beginning of the Great Leap Forward that China's industrial production would soon overtake that of Great Britain, for instance, it was actually Deng Xiaoping's Reform and Opening Up Campaign two decades later that ultimately enabled China's economy to surpass not only that of Great Britain but also that of every other country in the world. And while the original Great Leap Forward attempted to dissolve family bonds by encouraging more communitarian approaches, the socioeconomic reforms implemented under Deng Xiaoping have resulted in vast patterns of internal migration that have effectively split up tens of millions of families and separated many rural residents from their land. Deng's Reform and Opening Up Campaign, accordingly, ultimately managed to realize many of the objectives that Mao's original three-year Great Leap Forward failed to accomplish. Through the distinctive name of its protagonist, therefore, Liu Zhenyun's novel offers a salient reminder of the historical legacy of the Great Leap Forward, and of the Maoist era in general, while simultaneously underscoring the dark underbelly of the contemporary emphasis on economic growth.

For a more specific example of the significance of the protagonist's given name, we may begin with the title of the work. Although Liu Zhenyun's novel is narrated in the third person, the *title* of the work is emphatically in the first person with its translation "I am Liu Yuejin." Structured as a simple equation, the title affirms an equivalence between the proper name Liu Yuejin and the first-person pronoun *I* (or *wo*, in Chinese) (or, to be more precise, the title uses the linking verb *jiao* ["to call"] to connect its subject and predicate: *Wo jiao Liu Yuejin* ["I am called Liu Yuejin"]). The key to understanding the function of the protagonist's name in the work's title lies not in the name itself but rather in the title's use of the pronoun *wo*. In grammatical terms, a pronoun is considered a deictic—or what Roman Jacobson calls a "shifter"—meaning that the word's referent "shifts" depending on the specific context of the utterance. If the referent of a pronoun is necessarily informed by its context, however, Liu Zhenyun's title is actually profoundly ambiguous, insofar as the precise context of the phrase in the title is actually indeterminate. The phrase in question appears only in the title of the work and not anywhere in the main body of the text. While one might assume, somewhat tautologically, that the implicit speaker of the phrase is actually Liu Yuejin himself (since he effectively identifies himself as such), if we consider the environment within which the title appears, a more intriguing possibility presents itself. On both the novel's cover and title page, the title phrase actually *does* appear with additional contextual information that might identify it—namely, the phrase *Liu*

Zhenyun zhu ("[written] by Liu Zhenyun"). At one level, of course, this latter phrase may be taken simply as a statement of the novel's authorship, but it may also be read more literally as a claim of authorship over the first-person statement with which it is explicitly paired: *I am Liu Yuejin, written by Liu Zhenyun*—which is to say, *I, Liu Zhenyun, am Liu Yuejin*. Read in this way, the title—and, by extension, arguably even the entire novel—may be seen as a form of mediated autobiography. Like the fictional Liu Yuejin, Liu Zhenyun is surnamed Liu, is from Henan, is currently living in Beijing, and was even born in 1958 (the first year of the Great Leap Forward). Liu Zhenyun himself could be seen as the spectral double of his protagonist, and to the extent that the novel revolves around Liu Yuejin's attempts to reclaim his identity, the novel's title suggests that the work *itself* may be seen as figuratively doing the precise opposite—appropriating Liu Yuejin's story for the author's own purposes.

There is a similar use of pronouns in *Chengshi ye shi women de* (The city is also ours), a quasi-autobiographical novel by He Zhenzong, a migrant author who came to Guangdong in 1994 and sixteen years later wrote the narrative that would become *The City Is Also Ours* based on his experiences (He 2010). The work was originally serialized as a radio broadcast in 2009 and was published in book form the following year. Like *I Am Liu Yuejin*, *The City Is Also Ours* is about rural migrants and revolves around a father-son relationship, specifically the narrator's relationship with his father, who gave him the literary training that subsequently enabled him to compose the work itself. It turns out that the novel's title is borrowed from a poem the narrator wrote to his wife and infant son shortly after securing new residency permits for them in the city, in which he notes proudly that now "the city is also ours." While the meaning of this phrase is reasonably clear in the context of the narrative, it becomes strategically indeterminate when cited in the title of the novel, where it potentially refers either to the novel's protagonist and his son (as in the narrative) or perhaps to the author and his implied readers. Similarly ambiguous, meanwhile, is the meaning of the word "also" (*ye*) in the title. As with the title's first-person plural pronoun, the adverb "also" is tacitly inclusive of other claims to ownership—though these corollary claims are, in this instance, left unspecified. That is to say, the "also" in the title implicitly acknowledges that there are competing claims to ownership of a communal social space like a city and that claims to social identity are necessarily collective in nature, but it does so without specifying who else, exactly, is "also" entitled to make these claims.

I Am Liu Yuejin, meanwhile, concludes where it began, with Liu Yuejin finishing his lunch in a small restaurant. At this point, Liu has just returned to the town of Luoshui in an attempt to track down Pengju and retrieve the 60,000 yuan Liu Yuejin believes is rightfully his. In Luoshui, however, Liu Yuejin learns that Pengju and his girlfriend have already gone to Shanghai, where they hope to make their fortune. Convinced that Pengju and his girlfriend will merely waste the money that he still believes is rightfully his, Liu Yuejin decides to follow them. But, first, realizing that he has been traveling nonstop since leaving home, Liu Yuejin drops by a small restaurant to have a bowl of noodles. As he is finishing his meal, he looks up and sees Ju Li—the businessman's wife whose purse and flash drive had caused Liu Yuejin so much trouble. It turns out that Ju Li has followed Liu Yuejin to Luoshui from Beijing because she wants Liu's help retrieving another item that was in her stolen purse. She explains that while everyone else was searching for the hard drive, she is actually more interested in something else that was in the stolen purse: a small card, about half the size of a bank card, that she desperately needs back. In this way, Liu Yuejin once again finds himself back where he started—searching simultaneously for the 60,000 yuan from the divorce settlement (which is now in the hands of Pengju and his girlfriend) and for an abstract token of value contained in Ju Li's purse.

The parameters of this new pair of searches, introduced at the end of the novel, are left indeterminate. We are told little, for instance, with respect to what specifically Pengju plans to do with the settlement money, or whether Liu Yuejin is correct in his fears that Pengju will merely waste it. Similarly, apart from a variety of vague references to how someone's life is on the line, we are not told anything about the card or why precisely Ju Li needs to recover it—or even whether it contains digital data (like a bank card) or printed data (like Liu Yuejin's stolen photo ID). Instead, the only concrete information we are given about the card is that it has a picture of the legendary Monkey King Sun Wukong on the front. Sun Wukong is the simian protagonist of the classic Ming dynasty novel *Xiyouji* (Journey to the West), which describes the fantastic adventures of a group of Tang dynasty pilgrims traveling to "the West" to retrieve some Buddhist sutras. When, near the end of the novel, the pilgrims finally succeed in obtaining the sutras and are on their way home, they suddenly discover that the texts they have been given are actually completely blank. The pilgrims therefore return to the Buddhist patriarch, who informs them that the blank texts they received "are actually true, wordless scriptures," but adds that if the pilgrims are determined to have sutras with actual words, he would be willing to accommodate them (Yu 2012). Like the wordless sutras, the mysterious Sun Wukong card in *I Am Liu Yuejin* is presented as a paradigmatic

MacGuffin—an object of an urgent quest, the precise significance of which is never specified. The card is a symbol of information and value, though its contents are deliberately left unspecified and its significance is contingent on its position within a narrative that, as yet, remains unwritten.

Notes

1 Liu Zhenyun (2007: 4). All translations are my own, unless otherwise noted.
2 Confucius, *The Analects*, Book XIII, chapter 3.
3 Dave Lyons, "China's Golden Shield Project: Myths, Realities and Context," http://docslide.us/documents/dave-lyons-chinas-golden-shield-project.html.
4 The blogger and activist Michael Anti is one of the critics who have advanced this latter argument. See Magistad (2011).
5 The number of actual mobile phone users is lower, because many users have multiple accounts.
6 See "CNNIC Report: China Now Has 564 Million Internet Users, More Than Half Are on Weibo," www.techinasia.com/cnnic-report-2012-china-internet-users-weibo-stats/.
7 One study comparing native speakers of English in Hong Kong and the United States, for instance, has in fact found evidence that the Hong Kong–based cohort hold intuitions that are closer to what would be predicted by the descriptivist model, while the U.S.-based cohort treat names as behaving more causally (Mallon et al. 2004).
8 Published in 2007, *I Am Liu Yuejin* is clearly set in the contemporary period, but we are not told the precise year. We do know, however, that in the novel Liu Yuejin is forty years old, meaning that if he had been born in the first year of the Great Leap Forward, that would suggest that the novel's main narrative unfolds in 1998.

REFERENCES

Abbas, Ackbar. 1997. *Hong Kong: Culture and the Politics of Disappearance.* Minneapolis: University of Minnesota Press.

———. 2003. "Cinema, the City, and the Cinematic." In *Global Cities: Cinema, Architecture, and Urbanism in a Digital Age*, edited by Linda Krause and Patrice Petro, 142–156. New Brunswick: Rutgers University Press.

———. 2008. "Faking Globalization." In *Other Cities, Other Worlds: Urban Imaginaries in a Globalizing Age*, edited by Andreas Huyssen, 243–264. Durham: Duke University Press.

Abramson, Daniel, and Yu Qi. 2011. "'Urban-Rural Integration' in the Earthquake Zone: Sichuan's Post-Disaster Reconstruction and the Expansion of the Chengdu Metropole." *Pacific Affairs* 84, no. 3: 495–523.

Anagnost, Ann. 1997. *National Past-times: Narrative, Representation and Power in Modern China.* Durham: Duke University Press.

Ashiwa, Yoshiko, and David Wank. 2009. "Making Religion, Making the State in Modern China: An Introductory Essay." In *Making Religion, Making the State*, edited by Yoshiko Ashiwa and David Wank, 1–21. Stanford: Stanford University Press.

Augé, Marc. 1995. *Non-Places: Introduction to an Anthropology of Supermodernity.* Translated by John Howe. London: Verso.

Banks, Tony. 2003. "Property Rights Reform in Rangeland China: Dilemmas on the Road to the Household Ranch." *World Development* 31, no. 12: 2129–2142.

Barmé, Geremie, ed. 1996. *Shades of Mao: The Posthumous Cult of the Great Leader.* New York: M. E. Sharpe.

Bao Zonghao. 2003. "Minying keji qiye jia: zhishi jingji shidai de xin Shanghairen" [Technology entrepreneurs: The New Shanghai people of the information economy age]. In *Xin Shanghairen* [New Shanghai people], edited by Shanghai Zhengda yanjiu suo. Hong Kong: Sanlian shudian.

Baudrillard, Jean. 1994. *Simulacra and Simulation.* Ann Arbor: University of Michigan Press.

Bauer, Kenneth, and Yonten Nyima. 2010. "Laws and Regulations Impacting the Enclosure Movement on the Tibetan Plateau of China." *Himalaya* 30, nos. 1–2: 23–38.

Berlant, Lauren. 2011. *Cruel Optimism*. Durham: Duke University Press.
Berry, Chris, Fran Martin, and Audrey Yue, eds. 2003. *Mobile Cultures: New Media in Queer Asia*. Durham: Duke University Press.
Boyer, M. Christine. 1994. *The City of Collective Memory: Its Historical Imagery and Architectural Entertainments*. Cambridge: MIT Press.
Boym, Svetlana. 2001. *The Future of Nostalgia*. New York: Basic Books.
Braester, Yomi. 2010a. *Painting the City Red: Chinese Cinema and the Urban Contract*. Durham: Duke University Press.
———. 2010b. "Photography at Tiananmen: Pictorial Frames, Spatial Borders, and Ideological Matrixes." *positions: east asia cultures critique* 18, no. 3 (Winter 2010): 633–670.
———. 2013. "The Architecture of Utopia: From Rem Koolhaas's Scale Models to RMB City." In *The Spectacle and the City*, edited by Jeroen de Kloet and Lena Scheen, 59–73. Amsterdam: Amsterdam University Press.
Brenner, Neil. 2014. *Implosions/Explosions. Towards a Study of Planetary Urbanization*. Berlin: Jovis.
Brown, Philip H., and Yilin Xu. 2010. "Hydropower Development and Resettlement Policy on China's Nu River." *Journal of Contemporary China* 66, no. 19: 777–779.
Buchanan, Ian. 2000. *Michel de Certeau: Cultural Theorist*. London: Sage.
Butler, Judith. 1993. *Bodies That Matter: On the Discursive Limits of "Sex."* New York: Routledge.
Cao, Jianjun, Emily T. Yeh, Nicholas M. Holden, Yangyang Yang, and Guozhen Du. 2013. "The Effects of Enclosures and Land-Use Contracts on Rangeland Degradation on the Qinghai-Tibetan Plateau." *Journal of Arid Environments* 97: 3–8.
Cao Jinqing. 2003. "Fangfa yu gushi: Wo xinmu zhong de 'Shanghai ren.'" [Method and story: My view of *Shanghai People*]. In *Xin Shanghairen* [New Shanghai people], edited by Shanghai Zhengda yanjiu suo. Hong Kong: Sanlian shudian.
Caruth, Cathy. 2010. *Unclaimed Experience: Trauma, Narrative and History*. Baltimore: Johns Hopkins University Press.
Center for International Exchanges. 2005. "Jingwai Jieye Tongji Gongzuo Yiben Qingkuang yu Fenxi" [The basic facts and analyses of the statistics on overseas employment of 2004]. Unpublished report, Center for International Exchanges, Ministry of Labor and Social Security, Beijing.
Chen, Chih-Jou Jay. 2004. *Transforming Rural China: How Local Institutions Shape Property Rights in China*. London: Routledge Curzon.
Chen Sihe. 2003. "Shanghai ren, Shanghai wenhua he Shanghai de zhishi fenzi" [Shanghai people, Shanghai culture, and Shanghai intellectuals]. In *Xin Shanghairen* [New Shanghai people], edited by Shanghai Zhengda yanjiu suo. Hong Kong: Sanlian shudian.
Chen Xueming. 2003. "Nuxing: Shanghai ren renge jiegou zhong de yige zhiming ruodian" [Servility: A vital weakness in Shanghai people's character]. In *Xin Shanghairen* [New Shanghai people], edited by Shanghai Zhengda yanjiu suo. Hong Kong: Sanlian shudian.

———. 2009. "Shehui zhuyi shichang jingji shi renlei fazhan de fangxiang" [The socialist market economy is the direction of human development]. *Zhongguo shehui kexue yuan bao* 33.

Chien Shiuh-Shen. 2013. "Chinese Eco-Cities: A Perspective of Land-speculation-oriented Local Entrepreneurialism." *China Information* 27, no. 2: 173–196.

China Daily. 2011. "Nujiang Hydro Project Back on Agenda." Beijing, February 1.

China's International Contractors' Association. 2004. *Zhongguo duiwai laowu hezuo niandu baogao 2004* [Annual report on China's international labor collaboration]. December, Beijing. Not paginated.

Chou, Wah-shan. 2000. *Tongzhi: Politics of Same-Sex Eroticism in Chinese Societies*. New York: Haworth.

Chu, Julie. 2010. *Cosmologies of Credit: Transnational Mobility and the Politics of Destination in China*. Durham: Duke University Press.

Chu, Yun-han. 2011. "Sources of Regime Legitimacy and the Debate over the Chinese Model." *Asian Barometer: A Comparative Survey of Democracy, Governance and Development*, Figures 1a, 1b, 2a and 2b. Working paper series jointly published by Globalbarometer. No. 52, Taipei.

Clark, Marco. 2009. "Climbing the Mountain Within: Understanding Development Impacts and Overcoming Change in Southwest China." M.A. thesis, Oregon State University, Corvallis, OR.

Clough, Patricia, ed. 2007. *The Affective Turn: Theorizing the Social*. Durham: Duke University Press.

Comaroff, Jean, and John Comaroff. 1999. "Occult Economies and Violences of Abstraction: Spreading the Gospel of Prosperity." *American Ethnologist* 26, no. 2: 279–303.

———. 2000. "Millennial Capitalism: First Thoughts on a Second Coming." *Public Culture* 12, no. 2: 291–343.

Costello, Susan. 2008. "The Flow of Wealth in Golok Pastoralist Society: Towards an Assessment of Local Financial Resources for Economic Development." In *Tibetan Modernities: Notes from the Field on Cultural and Social Change*, edited by Robert Barnett and Ronald Schwartz, 73–112. Leiden: Brill.

C.S.M. 2013. "Chinese Online Literature: Voices in the Wilderness." *The Economist*. www.economist.com/blogs/prospero/2013/03/chinese-online-literature.

Cui, Zi'en. 2008. *Zhi tongzhi* [Queer China]. Beijing: dGenerate Films.

da Col, Giovanni. 2007. "The View from Somewhen: Events, Bodies and the Perspective of Fortune around Khawa Karpo, a Tibetan Sacred Mountain in Yunnan Province." *Inner Asia* 9, no. 2: 215–235.

Dai Shenzhi. 2012. "Learning from the 2008 Wenchuan Earthquake's Recovery and Reconstruction Planning." Presented at the University of North Carolina at Chapel Hill, March 26, 2012.

Dai Zhikang. 2003a. "Dazao da Shanghai" [Building up Greater Shanghai]. In *Xin Shanghairen* [New Shanghai people], edited by Shanghai Zhengda yanjiu suo. Hong Kong: Sanlian shudian.

———. 2003b. "Qianyan" [Preface]. In *Xin Shanghairen* [New Shanghai people], edited by Shanghai Zhengda yanjiu suo. Hong Kong: Sanlian shudian.

de Certeau, Michel. 1984. "Walking in the City." In *The Practice of Everyday Life*, vol. 1, translated by Steven Rendall, 91–110. Berkeley: University of California Press.

———. 1998. "Ghosts in the City." In Michel de Certeau, Luce Giard, and Pierre Mayol, *Practice of Everyday Life, Volume Two: Living and Cooking*, translated by Timothy J. Tomasik, 133–143. Minneapolis: University of Minnesota Press.

De Flander, Katleen. 2015. "Closed Cycles, Open Cities." In *The Urban Climate Challenge: Rethinking the Role of Cities in the Global Climate Regime*, edited by Craig Johnson, Noah Toly, and Heike Shroeder, 37–62. New York: Routledge.

De Soto, Hernando. 2003. *The Mystery of Capital: Why Capitalism Triumphs in the West and Fails Everywhere Else*. New York: Basic Books.

Deng Xi. 2012. "Dujiangyan lüyou pinpai de jiazhi xiaoying yanjiu" [Study of the effective value of Dujiangyan's tourism brand]. *Quyu Jingji* [Regional Economics] 16: 137–138.

Deng Weizhi. 2003. "Wangmin shi xin Shanghairen de zhuti" [Netizens are the subjects of New Shanghai]. In *Xin Shanghairen* [New Shanghai people], edited by Shanghai Zhengda yanjiu suo. Hong Kong: Sanlian shudian.

Derrida, Jacques. 1994. *Specters of Marx: The State of the Debt, the Work of Mourning, and the New International*. New York: Routledge.

———. 1998. *Of Grammatology*. Baltimore: Johns Hopkins University Press.

Des Forges, Alexander. 2007. *Mediasphere Shanghai: The Aesthetics of Cultural Production*. Honolulu: University of Hawai'i Press.

Dong, Madeleine Yue. 2003. *Republican Beijing: The City and Its Histories*. Berkeley: University of California Press.

Duan Jin, and Yin Ming. 2011. *Dangdai xincheng kongjian fazhan yanhua guilü anlie genzong yanjiu yu weilai guihua sikao* [Case studies of the spatial evolution of contemporary New Towns and thoughts on future planning]. Nanjing: Dongnan University Press.

Dunch, Ryan. 2008. "Christianity and 'Adaptation to Socialism.'" In *Chinese Religiosities: Afflictions of Modernity and State Formation*, edited by Mayfair Yang, 165–188. Berkeley: University of California Press.

Economy, Elizabeth C. 2004. *The River Runs Black: The Environmental Challenge to China's Future*. Ithaca: Cornell University Press.

Edelman, Marc. 2005. "Bringing the Moral Economy Back in . . . to the Study of 21st Century Transnational Social Movements." *American Anthropologist* 107, no. 3: 331–345.

Ellingson, Stephen. 1995. "Understanding the Dialectic of Discourse and Collective Action: Public Debate and Rioting in Antebellum Cincinnati." *American Journal of Sociology* 101: 100–144.

Ethnologue. 2009. "Lisu: A Language of China." www.ethnologue.com.

Faier, Lieba, and Lisa Rofel. 2014. "Ethnographies of Encounter." *Annual Review of Anthropology* 43: 363–377.

Fang, Gang. 1995. *Tongxinglian zai Zhongguo* [Homosexuality in China]. Changchun: Jilin People's Press.

Feng Deng. 2010. "Post-Disaster Reconfiguration of Property Rights in a Transitional Economy." *Post-Communist Economies* 22, no. 2: 193–206.

Flower, John. 2009. "Ecological Engineering on the Sichuan Frontier: Socialism as Development Policy, Local Practice, and Contested Ideology." *Social Anthropology* 17, no. 1: 40–55.

Ford, Jocelyn. 2014. "Prejudice, Exclusion and Sexism Is All Part of Life for a Tibetan Migrant in Beijing." www.pri.org/stories/2014-08-28/new-documentary-shows-hidden-sexism-ridden-lives-tibetan-women.

Frege, Gottlob. 1952. "On Sense and Reference." In *Translations from the Philosophical Writings of Gottlob Frege*, edited by P. Geach and M. Black. Oxford: Blackwell.

Freud, Sigmund. 1920. *Beyond the Pleasure Principle*. Vol. 18 in The Standard Edition, edited and translated by James Strachey et al. London: Hogarth, 1953–1974.

———. 1937–1939. *Moses and Monotheism*. The Standard Edition, vol. 23, edited and translated by James Strachey. New York: Vintage.

———. 1959. "A Note Upon the 'Mystic Writing Pad.'" In *Collected Papers: Volume 5 (Miscellaneous Papers, 1888–1938)*, edited by James Strachey. London: Basic Books.

———. 1962. *Civilization and Its Discontents*, translated by James Strachey. New York: W. W. Norton.

Friedberg, Anne. 2004. "Virilio's Screen: The Work of Metaphor in the Age of Technological Convergence." *Journal of Visual Culture* 3, no. 2: 183–193.

Friedman, Eli. 2013. "Outside the New China." *Jacobin*, 11–12. www.jacobinmag.com/2013/09/outside-the-new-china/.

———. 2014. *Insurgency Trap: Labor Politics in Postsocialist China*. Ithaca: Cornell University Press.

Gaerrang. 2012. "Alternative Development on the Tibetan Plateau: The Case of the Slaughter Renunciation Movement." PhD dissertation. University of Colorado, Boulder.

Galipeau, Brendan A., Mark Ingman, and Bryan Tilt. 2013. "Dam-Induced Displacement and Agricultural Livelihoods in China's Mekong Basin." *Human Ecology* 41: 437–446.

Gao, Boyang, Liu Weidong, and Michael Dunford. 2014. "State Land Policy, Land Markets and Geographies of Manufacturing: The Case of Beijing, China." *Land Use Policy* 36: 1–12.

Gaubatz, Piper. 2008. "Commercial Redevelopment and Regional Inequality in Urban China: Xining's Wangfujing." *Eurasian Geography and Economics* 49, no. 2: 180–199.

Gayley, Holly. 2011. "The Ethics of Cultural Survival: A Buddhist Vision of Progress in Mkhan po 'Jigs phun's Heart Advice to Tibetans for the 21st Century." In

Mapping the Modern in Tibet, edited by Gray Tuttle, 435–502. Proceedings of the Eleventh Seminar of the International Association for Tibetan Studies.

Germano, David. 1998. "Re-membering the Dismembered Body of Tibet: Contemporary Tibetan Visionary Movements in the People's Republic of China." In *Buddhism in Contemporary Tibet: Religious Revival and Cultural Identity*, edited by Melvyn Goldstein and Matthew Kapstein, 53–94. Berkeley: University of California Press.

Gong Hui. 2009. "Considerations Regarding Legal Protection for Chinese Homosexuals." *Journal of Handan Polytechnic College* 22, no. 3: 14–16. http://d.wanfangdata.com.cn/periodical_hdzyjsxyxb200903004.aspx.

Grace, Helen. 2007. "Monuments and the Face of Time: Distortions of Scale and Asynchrony in Postcolonial Hong Kong." *Postcolonial Studies* 10, no. 4: 467–483.

Graeber, David. 2001. *Toward an Anthropological Theory of Value: The False Coin of Our Own Dreams*. New York: Palgrave.

Griffith, David. 2009. "The Moral Economy of Tobacco." *American Anthropologist* 111, no. 4: 432–442.

Grube, Katherine. 2009. "Ai Weiwei Challenges China's Government over Earthquake." *ArtAsiaPacific*, July/August. http://artasiapacific.com/Magazine/64/AiWeiweiChallengesChinasGovernmentOverEarthquake.

Grumbine, Edward R. 2010. *Where the Dragon Meets the Angry River: Nature and Power in the People's Republic of China*. Washington, DC: Island.

Gu Xiaoming. 2003. "'Shanghai ren' de zulei neihan he 'Xin Shanghairen'" [The in-group connotations of "Shanghai People" and "New Shanghai people"]. In *Xin Shanghairen* [New Shanghai people], edited by Shanghai Zhengda yanjiu suo. Hong Kong: Sanlian shudian.

Guo, Ling, et al. 2012. "Research of Tourism Showtime Products in the Urban Tourism Development." *Proceedings of the 2nd International Conference on Green Communications and Networks 2012* (GCN 2012) 3, no. 225: 421–427.

Guo, Yan. 2012. "Urban Resilience in Post-Disaster Reconstruction: Towards a Resilient Development in Sichuan, China." *International Journal of Disaster Risk Science* 3, no. 1: 45–55.

Hankiss, Elemér. 1988. "The 'Second Society': Is There an Alternative Social Model Emerging in Contemporary Hungary?" *Social Research* 55, no. 1: 13–42.

Hardt, Michael. 1999. "Affective Labor." *boundary 2* 26, no. 2: 89–100.

Harrell, Stevan. 1995. "Introduction: Civilizing Projects and the Reaction to Them." In *Cultural Encounters on China's Ethnic Frontiers*, edited by Stevan Harrell, 3–36. Seattle: University of Washington Press.

Harris, Richard. 2010. "Rangeland Degradation on the Qinghai-Tibetan Plateau: A Review of the Evidence of Its Magnitude and Causes." *Journal of Arid Environments* 74: 1–12.

Harvey, David. 2000a. *Megacities Lecture 4*. Amersfoort: Twynstra Gudde Management Consultants.

———. 2000b. *Spaces of Hope*. Berkeley: University of California Press.

———. 2005. *A Brief History of Neo-liberalism*. New York: Oxford University Press.

———. 2009. *Social Justice and the City*. Athens: University of Georgia Press.

He, Daming, Yan Feng, and Jinming Hu. 2007. *Zhongguo xinan guoji heliu shui ziyuan liyong yu shengtai baohu* [Utilization of water resources and environmental conservation in the international rivers, southwest China]. Beijing: Science Press.

He Ge. 2012. "Evaluation on Land Use Planning Based on Ecosystem Services Value Change: A Case Study of Dujiangyan City." *Advances in Natural Science* 5, no. 2: 41–44.

He Zhenzong. 2010. *Chengshi ye shi women de: Yi ge nongmingong zai Guangdong 18 nian de fendou shi* [The city is also ours: A rural migrant worker's eighteen years of struggle]. Yinchuan: Ningxia renmin chubanshe.

Heidegger, Martin. (1954) 2008. "The Question Concerning Technology." In *Basic Writings*, ed. David Farrell Krell, 307–342. New York: HarperCollins.

Hershatter, Gail. 2011. *The Gender of Memory: Rural Women and China's Collective Past*. Berkeley: University of California Press.

Ho, Loretta. 2010. *Gay and Lesbian Subculture in Urban China*. London: Routledge.

Ho, Peter. 2000. "The Clash over State and Collective Property: The Making of the Rangeland Law." *China Quarterly* 161: 240–263.

———. 2001. "Who Owns China's Land? Policies, Property Rights and Deliberate Institutional Ambiguity." *China Quarterly* 166: 394–421.

Hsing, You-tien. 2010. *The Great Urban Transformation: Politics of Land and Property in China*. New York: Oxford University Press.

Hu Shouyun. 2003. "Xin Shanghairen ying jubei naxie yishi?" [Which consciousnesses should New Shanghai people possess?]. In *Xin Shanghairen* [New Shanghai people], edited by Shanghai Zhengda yanjiu suo. Hong Kong: Sanlian shudian.

Huang, Tsung-yi Michelle. 2004. *Walking between Slums and Skyscrapers: Illusions of Open Space in Hong Kong, Tokyo, and Shanghai*. Hong Kong: Hong Kong University Press.

Huang, Youqing, and Si-ming Li. 2014. *Housing Inequality in Chinese Cities*. London: Routledge.

Hult, Anna. 2013. "Swedish Production of Sustainable Urban Imaginaries in China." *Journal of Urban Technology* 20, no. 1: 77–94.

Human Rights in China. 2002. "Shutting out the Poorest: Discrimination against the Most Disadvantaged Migrant Children in City Schools."

Hung, Ho-fung. 2013. "Labor Politics under Three Stages of Chinese Capitalism." *South Atlantic Quarterly* 112, no. 1: 203–212.

Huyssen, Andreas. 2003. *Present Pasts: Urban Palimpsests and the Politics of Memory*. Stanford: Stanford University Press.

International Rivers. 2009. "Salween Dams." http://internationalrivers.org/en/southeast-asia/burma/salween-dams.

———. 2013. "Spreadsheet of Major Dams in China." Berkeley: International Rivers.

Jameson, Fredric. 1991. *Postmodernism, or, the Cultural Logic of Late Capitalism.* Durham: Duke University Press.

Jin, Yuhui. 2014. "One City, Nine 'Ghost Towns': Seeking an Alternative Growth Model to Revitalize Lingang New Harbor City in Shanghai." Graduate poster, TU Delft Department of Urbanism, June 23.

Kang Yanhong. 2006. "Zhengfu qiye beijing xia de Zhongguo chengshi jiaoqushi fazhang yanjiu" [Chinese suburbanization development in the context of government entrepreneurialism]. *Renmin dili* [Human geography] 21, no. 5: 10–13.

Keil, Roger. 2013. *Suburban Constellations: Governance, Land and Infrastructure in the 21st Century.* Berlin: Jovis Verlag.

Kim, Jung In. 2014. "Making Cities Global: The New City Development of Songdo, Yujiapu, and Lingang." *Planning Perspectives* 29, no. 3: 329–356.

Kirby, William, ed. 2011. *The People's Republic of China at 60: An International Assessment.* Cambridge: Harvard University Asia Center.

Kochan, Dror. 2009. "Visual Representation of Internal Migration and Social Change in China." *China Information* 23, no. 2: 285–316.

Kojève, Alexander. 1962. *Introduction à la lecture de Hegel: Leçons sur "La Phénoménologie de l'Espirit."* 2nd ed. Paris: Gallimard.

Kong, Shuyu. 2005. *Consuming Literature: Best Sellers and the Commercialization of Literary Production in Contemporary China.* Stanford: Stanford University Press.

Kong, Travis S. K. 2011. *Chinese Male Homosexualities: Memba, Tongzhi, and Golden Boy.* New York: Routledge.

Kripke, Saul. 1980. *Naming and Necessity.* Cambridge: Harvard University Press.

Kristeva, Julia. 1982. *Powers of Horror: An Essay on Abjection.* Translated by Leon S. Roudiez. New York: Columbia University Press.

Kwong, Julia. 2004. "Educating Migrant Children: Negotiations between the State and Civil Society." *China Quarterly* 180: 1073–1088.

Lacan, Jacques. (1949) 2002. *Ecrits.* Translated by Bruce Fink. New York: W. W. Norton.

Lawrence, Regina G. 2001. "Defining Events: Problem Definition in the Media Arena." In *Politics, Discourse, and American Society: New Agendas,* edited by Roderick P. Hart and Bartholomew H. Sparrow, 91–110. Lanham, MD: Rowman and Littlefield.

Lee, Ching Kwan. 1998. *Gender and the South China Miracle.* Berkeley: University of California Press.

———. 2007. *Against the Law: Labor Protests in China's Rustbelt and Sunbelt.* Berkeley: University of California Press.

Lefebvre, Henri. 2003. *The Urban Revolution.* Translated by Robert Bononno. Minneapolis: University of Minnesota Press.

Li, Lianjiang. 2004. "Political Trust in Rural China." *Modern China* 30, no. 2: 228–258.

Li, Miao. 2015. *Citizenship Education and Migrant Youth in China: Pathways to the Urban Underclass*. New York: Routledge.

Li, Qiang, and Satish Chand. 2013. "Evidence of a Housing Bubble in Beijing." *Academic Journal of Interdisciplinary Studies* 2, no. 8: 627–634.

Li, Tania. 2007. *The Will to Improve: Development, Governmentality, and the Practice of Politics*. Durham: Duke University Press.

Li, Xinran. 2008. "The Benefits of Hydropower Development on the Nu River for Increasing Local Farmers' Incomes." In *Nujiang, Lancangjiang and Jinshajiang: Research on the Exploitation of Hydropower Resources and Environmental Protection*, edited by Yaohua He, 264–279. Kunming: Southwest Nationalities Research.

Li Zhen. 2014. "Shanghai Lingang diqu chanye jiju yu chengshihua fazhan hudong guanxi yanjiu" [Study on the interaction between industry and city development of Lingang in Shanghai]. *Shanghai chengshi guice* [Shanghai Urban Planning Review] 1: 122–125.

Li Zhiming. 2012. "Study on the 'Housing with Limited Property Rights' in Transitional China: A Case of Xingwei Village in Nanjing, China." Presented at the conference proceedings of the China Planning Conference IACP.

Liang, Samuel Y. 2010. "Property-Driven Urban Change in Post-Socialist Shanghai: Reading the Television Series *Woju*." *Journal of Current Chinese Affairs* 39, no. 4: 3–28.

Lin, Chun. 2006. *The Transformation of Chinese Socialism*. Durham: Duke University Press.

Lin, Tuo, et al. 2011. "The Demonstration Effects and Potential Problems of Chinese Marine Eco-City on Environment, Energy and Planning." Presented at the Artificial Intelligence, Management Science and Electronic Commerce (AIMSEC) 2nd International Conference.

Litzinger, Ralph. 2007. "In Search of the Grassroots: Hydroelectric Politics in Northwest Yunnan." In *Grassroots Political Reform in Contemporary China*, edited by Elizabeth J. Perry and Merle Goldman, 282–299. Cambridge: Harvard University Press.

———, ed. 2013. "Labor in China: A New Politics of Struggle." Special issue of *South Atlantic Quarterly* 112, no. 1: 172–212.

———. Forthcoming. *Migrant Futures: China from the Urban Fringe*.

Liu, Zhenya. 2012. *Zhongguo dianli yu nengyuan* [Electric power and energy in China]. Beijing: China Electric Power.

Liu Zhenyun. 2007. *Wo jiao Liu Yuejin* [I am Liu Yuejin]. Wuhan: Changjiang wenyi chubanshe.

Lu, Hanchao. 2002. "Nostalgia for the Future: The Resurgence of an Alienated Culture in China." *Pacific Affairs* 75, no. 2: 169–186.

Lu, Xiuzhang. 2008. "On the Relationship between Dam Construction, Environmental Protection and Ethnic Minorities in the Nu River Basin." In *Nujiang, Lancangjiang and Jinshajiang: Research on the Exploitation of Hydropower*

Resources and Environmental Protection, edited by Yaohua He, 101–107. Kunming: Southwest Nationalities Research.

Lugg, Alexander. 2011. "Chinese Online Fiction: Taste Publics, Entertainment, and Candle in the Tomb." Chinese Journal of Communication 4, no. 2: 121–136.

Luo Tingting, and Gao Xiaowei. 2011. "New Urban Development of Beijing's Real Estate Development Level of the Fuzzy Clustering Analysis." IEEE: 500–504.

Magee, Darrin. 2006. "Powershed Politics: Hydropower and Interprovincial Relations under Great Western Development." China Quarterly 185: 23–41.

Magee, Darrin, and Kristen McDonald. 2009. "Beyond Three Gorges: Nu River Hydropower and Energy Decision Politics in China." Asian Geographer 25, nos. 1–2: 39–60.

Magistad, Mary Kay. 2011. "Chinese Online Anger over Train Tragedy." PRI's The World, July 29. www.theworld.org/2011/07/chinese-online-anger-over-train-tragedy/.

Makley, Charlene. 2014. "The Amoral Other: State-led Development and Mountain Deity Cults among Tibetans in Amdo Rebgong." In Mapping Shangrila: Contested Landscapes in the Sino-Tibetan Borderlands, edited by Emily T. Yeh and Chris Coggins, 229–254. Seattle: University of Washington Press.

Mallon, Ron, et al. 2004. "Semantics Cross-Cultural Style." Cognition 92: B1–B12.

Manovich, Lev. 2001. The Language of New Media. Cambridge: MIT Press.

Marx, Karl. 1977. Capital, vol. 1. Translated by Ben Fowkes. New York: Vintage.

———. (1867) 1978. Capital, Vol. 1. In The Marx-Engels Reader, 2nd ed., edited by Robert C. Tucker. New York: W. W. Norton.

Marx, Karl, and Friedrich Engels. 1998. Manifesto of the Communist Party. New York: Merlin.

Mattern, Shannon. 2008. "Broadcasting Space: China Central Television's New Headquarters." International Journal of Communication 2: 869–908.

McDonagh, Josephine. 1987. "Writings on the Mind: Thomas De Quincey and the Importance of the Palimpsest in Nineteenth Century Thought." Prose Studies 10, no. 2: 207–224.

McDougall, Bonnie S. 2003. Fictional Authors, Imaginary Audiences: Modern Chinese Literature in the Twentieth Century. Hong Kong: Chinese University Press.

McGrath, Jason. 2008. Postsocialist Modernity: Chinese Cinema, Literature, and Criticism in the Market Age. Stanford: Stanford University Press.

McMahan, David. 2008. The Making of Buddhist Modernism. Oxford: Oxford University Press.

Meng Yue. 2006. Shanghai and the Edges of Empire. Minneapolis: University of Minnesota Press.

Merrifield, Andy. 2014. The New Urban Question. London: Pluto Press.

Mertha, Andrew C. 2008. China's Water Warriors: Citizen Action and Policy Change. Ithaca: Cornell University Press.

Meyer, Birgit. 1998. "Commodities and the Power of Prayer: Pentecostalist Attitudes towards Consumption in Contemporary Ghana." Development and Change 29, no. 4: 751–776.

Ming, Holly. 2014. *The Education of Migrant Children and China's Future: The Urban Left Behind*. London: Routledge.

Ministry of Commerce. 2012. "2011 nian woguo duiwai laowu hezuo yewu jianmin tongji" [Summary statistics on China's international labor cooperation 2011]. http://fec.mofcom.gov.cn/article/tjzl/lwhz/201201/1276088_1.html.

Miyazaki, Hirokazu. 2004. *The Method of Hope: Anthropology, Philosophy, and Fijian Knowledge*. Stanford: Stanford University Press.

Moore, Sally. 1987. "Explaining the Present: Theoretical Dilemmas in Processual Ethnography." *American Ethnologist* 14: 727–736.

Mountford, Tom. 2010. "The Legal Status and Position of Lesbian, Gay, Bisexual and Transgender People in the People's Republic of China." International Gay and Lesbian Human Rights Commission. www.iglhrc.org/cgi-bin/iowa/article/takeaction/resourcecenter/1107.html.

Naftali, Orna. 2014. *Children, Rights, and Modernity in China: Raising Self-Governing Citizens*. New York: Palgrave.

Oliver-Smith, Anthony. 2010. *Development and Dispossession: The Crisis of Forced Displacement and Resettlement*. Santa Fe, NM: School for Advanced Research Press.

Ong, Aihwa. 1997. "Chinese Modernities: Narratives of Nation and of Capitalism." In *Ungrounded Empires*, edited by Ong and Nonini. London: Routledge.

Ong, Aihwa. 2006. *Neoliberalism as Exception: Mutations in Citizenship and Sovereignty*. Durham: Duke University Press.

Ong, Aihwa, and Donald Nonini, eds. 1997. *Ungrounded Empires: The Cultural Politics of Chinese Transnationalism*. London: Routledge.

Ong, Aihwa, and Li Zhang. 2008. *Privatizing China: Socialism from Afar*. Ithaca: Cornell University Press.

Pai, Hsien-Yong. (1983) 1990. *Niezi* [Crystal boys]. Translated by Howard Goldblatt. San Francisco: Gay Sunshine Press.

Paik, Wooyeal, and Kihyun Lee. 2012. "I Want to Be Expropriated!: The Politics of *Xiaochanquanfang* Land Development in Surburban China." *Journal of Contemporary China* 21, no. 74: 261–279.

Parke, Elizabeth. 2015. "Migrant Workers and the Imagining of Human Infrastructure in Chinese Contemporary Art." *China Information* 29, no. 2: 226–252.

Peng Xuan. 2014. "Tongzhou xincheng jianshe jinzhan yanjiu—jiyu xincheng gongneng jiaodu de fenxi" [Study of Tongzhou New City construction progress—an analysis from the perspective of New City functions]. *Xiandai Shanghai* [Modern Shanghai]: 168–169.

Phuntso, Karma. H.H. 2004. "Khenpo Jigme Phuntso: A Tribute and a Translation." *Journal of Bhutan Studies*, no. 11: 129–136.

Pong, Myra. 2015. *Educating the Children of Migrant Workers in Beijing: Migration, Education, and Policy in Urban China*. London: Routledge.

Pun, Ngai. 2005. *Made in China: Women Factory Workers in a Global Workplace*. Durham: Duke University Press.

Pun, Ngai, and Jenny Chan. 2013. "The Spatial Politics of Labor in China: Life, Labor, and a New Generation of Migrant Workers." *South Atlantic Quarterly* 112, no. 1: 179–190.

Queen, Christopher S., and Sallie B. King. 1996. *Engaged Buddhism: Buddhist Liberation Movements in Asia.* Albany: State University of New York Press.

Ribot, Jesse C., and Nancy Lee Peluso. 2003. "A Theory of Access." *Rural Sociology* 68, no. 2: 153–181.

Rofel, Lisa. 1999. *Other Modernities: Gendered Yearnings in China after Socialism.* Berkeley: University of California Press.

———. 2007. *Desiring China: Experiments in Neoliberalism, Sexuality and Public Culture.* Durham: Duke University Press.

———. 2012. "Between Tianxia and Postsocialism: Contemporary Chinese Cosmopolitanism." In *Routledge International Handbook of Cosmopolitan Studies*, edited by Gerard Delanty, 443–451. London: Routledge.

———. 2014. "Transnational Business between China and Italy in the Fashion Industry." Unpublished talk given at Duke University.

Rofel, Lisa, and Sylvia Yanagisako. Forthcoming. *The Twenty-First-Century Silk Road: Transnational Capitalism between China and Italy in the Fashion Industry.* Durham: Duke University Press.

Rose, Nicholas. 1999. *Powers of Freedom: Reframing Political Thought.* Cambridge: Cambridge University Press.

Rossabi, Morris. 2004. "Introduction." In *Governing China's Multiethnic Frontiers*, edited by Morris Rossabi, 3–18. Seattle: University of Washington Press.

Roy, Ananya. 2009. "The 21st Century Metropolis: New Geographies of Theory." *Regional Studies* 43, no. 6: 819–830.

Roy, Ananya, and Nezar AlSayyad, eds. 2003. *Urban Informality: Transnational Perspectives from the Middle East, Latin America, and South Asia.* Lanham, MD: Lexington.

Rozelle, Scott, Loren Brandt, Li Guo, and Jikun Huang. 2005. "Land Tenure in China: Facts, Fictions and Issues." In *Developmental Dilemmas: Land Reform and Institutional Change in China*, edited by Peter Ho, 121–150. New York: Routledge.

Rudnyckyj, Daromir. 2008. "Worshipping Work: Producing Commodity Producers in Contemporary Indonesia." In *Taking Southeast Asia to Market: Commodities, Nature and People in the Neoliberal Age*, edited by Joseph Nevins and Nancy Lee Peluso, 73–87. Ithaca: Cornell University Press.

———. 2009. "Spiritual Economies: Islam and Neoliberalism in Contemporary Indonesia." *Cultural Anthropology* 24, no. 1: 104–141.

Rudolph, Maya Eva Gunst. 2012. "CinemaTalk: A Conversation with Filmmaker Ji Dan." http://dgeneratefilms.com/critical-essays/cinematalk-a-conversation-with-ji-dan.

Russell, Bertrand. 1956. "On Denoting." Reprinted in *Logic and Knowledge*, edited by R. C. Marsh. London: George Allen and Unwin.

Rutherford, Danilyn. 2003. *Raiding the Land of the Foreigners.* Princeton: Princeton University Press.

Sahlins, Marshall. 1976. *Culture and Practical Reason.* Chicago: University of Chicago Press.

———. 1985. *Islands of History.* Chicago: University of Chicago Press.

Scott, James C. 1976. *The Moral Economy of the Peasant.* New Haven: Yale University Press.

———. 1990. *Domination and the Arts of Resistance: Hidden Transcripts.* New Haven: Yale University Press.

Scotton, Carol M., and Wanjin Zhu. 1983. "Tongzhi in China: Language Change and Its Conversational Consequences." *Language in Society* 12: 477–494.

Scudder, Thayer. 2005. *The Future of Large Dams: Dealing with Social, Environmental, Institutional and Political Costs.* London: Earthscan.

Sewell, William H. 2005. *Logics of History: Social Theory and Social Transformation.* Chicago: University of Chicago Press.

Sha, Yongjie, et al. 2014. *Shanghai Urbanism at the Medium Scale.* Berlin: Springer-Verlag.

Shao, Qin. 2013. *Shanghai Gone: Domicide and Defiance in a Chinese Megacity.* Plymouth, MA: Rowman and Littlefield.

Shanghai Zhengda yanjiu suo, ed. 2003. *Xin Shanghairen* [New Shanghai people]. Hong Kong: Sanlian shudian.

Shen, Jie, and Fulong Wu. 2013. "Moving to the Suburbs: Demand-Side Driving Forces of Suburban Growth in China." *Environment and Planning* 45: 1823–1844.

Shi, Jiangtao. 2009. "Wen Calls Halt to Yunnan Dam Plans." *South China Morning Post.* Beijing, May 21.

Shi Runjiu, dir. 1999. *Meili xin shijie* [Beautiful new world]. Xi'an Film Studio.

Shue, Vivienne. 2004. "Legitimacy Crisis in China?" In *State and Society in 21st-Century China*, edited by Peter Hays Gries and Stanley Rosen, 24–49. New York: Routledge Curzon.

Sieber, Patricia. 2001. *Red Is Not The Only Color: Contemporary Chinese Fiction on Love and Sex between Women, Collected Stories*, edited by Patricia Sieber, 1–35. Lanham, MD: Rowman and Littlefield.

Siegel, Greg. 2002. "Double Vision: Large-Screen Video Display and Live Sports Spectacle." *Television and New Media* 3, no. 1: 49–73.

Sigley, Gary. 2006. "Sex, Politics and the Policing of Virtue in the People's Republic of China." In *Sex and Sexuality in China*, edited by Elaine Jeffreys, 43–61. London: Routledge.

Sivaramakrishnan, Kalyanakrishnan. 2005. "Introduction to 'Moral Economies, State Spaces, and Categorical Violence.'" *American Anthropologist* 107, no. 3: 321–330.

Smil, Vaclav. 2004. *China's Past, China's Future: Energy, Food, Environment.* New York: Routledge.

Smith, Nick. 2014. "Living on the Edge: Household Registration Reform and Peri-Urban Precarity in China." *Journal of Urban Affairs* 36, no. S1: 369–383.

So, Alvin Y. 2007. "Peasant Conflict and the Local Predatory State in the Chinese Countryside." *Journal of Peasant Studies* 34, nos. 3–4: 560–581.

Soja, Edward W. 1997. "Six Discourses on the Postmetropolis." In *Imagining Cities: Scripts, Signs, Memory*, edited by Sallie Westwood and John Williams, 19–30. London: Routledge.

———. 2000. *Postmetropolis: Critical Studies of Cities and Regions*. Oxford: Blackwell.

Solinger, Dorothy. 2013. "Temporality as Trope in Delineating Inequality: Progress for the Prosperous, Time Warp for the Poor." In *Unequal China: The Political Economy and Cultural Politics of Inequality*, edited by Wanning Sun and Yingjie Guo, 59–76. London: Routledge.

Sorace, Christian. 2014. "China's Last Communist: Ai Weiwei." *Critical Inquiry* 40, no. 2: 396–419.

State Council. 2006. "Da zhong xing shuili shuidian gongcheng jianshe zhengdi buchang he yimin anzhi tiaoli" [Regulations on land acquisition compensation and resettlement of migrants for construction of large and medium scale water conservancy and hydropower projects].

Steinhardt, H. Christoph. 2012. "Contesting 'Social Stability': Understanding the Dynamics of Grassroots Activism and Intellectual Critique in Contemporary China." Presented at the conference Modes of Activism and Engagement in the Chinese Public Sphere, April 26–27, 2012, organized by Asia Research Institute, National University of Singapore.

Sun, Wanning. 2014. *Subaltern China: Rural Migrants, Media, and Cultural Practices*. Lanham, MD: Rowman and Littlefield.

Swearer, Donald. 1996. "Sulak Sivaraksa's Buddhist Vision for Renewing Society." In *Engaged Buddhism: Buddhist Liberation Movements in Asia*, edited by Christopher S. Queen and Sallie B. King, 217–218. Albany: State University of New York Press.

Sze, Julie. 2015. *Fantasy Islands: Chinese Dreams and Ecological Fears in an Age of Climate Crisis*. Oakland: University of California Press.

Tan, Xuewen. 2010. "New-Town Development and Policy in China." *Chinese Economy* 43, no. 3: 47–58.

Tang Daqian. 2008. "Yi gonggong jiaotong wei daoxiang de TOD xincheng yanjiu" [A study of transportation-oriented development of New Towns]. Master's thesis, Tianjin University.

Taussig, Michael. 1983. *The Devil and Commodity Fetishism in South America*. Chapel Hill: University of North Carolina Press.

Thompson, Edward P. 1971. "The Moral Economy of the English Crowd in the 18th Century." *Past and Present* 50: 76–136.

Tilt, Bryan. 2010. *The Struggle for Sustainability in Rural China: Environmental Values and Civil Society*. New York: Columbia University Press.

———. 2015. *Dams and Development in China: The Moral Economy of Water and Power.* New York: Columbia University Press.

Tsafrir, Yoram, Shmuel Safrai, and Menahem Stern, eds. 1999. *Sefer Yrushalayim: Ha-tkufah ha-romit veha-bizantit* [The book of Jerusalem: The Roman and Byzantine period]. Jerusalem: Yad Ben-Zvi.

Tullos, Desiree D., Eric Foster-Moore, Darrin Magee, Bryan Tilt, Aaron T. Wolf, Edwin Schmitt, Francis Gassert, and Kelly Kibler. 2013. "Biophysical, Socioeconomic and Geopolitical Vulnerabilities to Hydropower Development on the Nu River, China." *Ecology and Society* 18, no. 3: 16–34.

Turner, Victor. 1974. *Dramas, Fields, and Metaphors: Symbolic Action in Human Society.* Ithaca: Cornell University Press.

Tweedie, James, and Yomi Braester. 2010. "Introduction: The City's Edge." In *Cinema at the City's Edge: Film and Urban Networks in East Asia*, edited by Yomi Braester and James Tweedie, 1–16. Hong Kong: Hong Kong University Press.

United Nations Development Programme (UNDP). 2008. "Human Development Report: China." Beijing: UNDP.

van der Werff, Ties. 2010. "The Struggle of the *Tongzhi* Homosexuality in China and the Position of Chinese 'Comrades.'" In *Urgency Required: Gay and Lesbian Rights Are Human Rights*, edited by Ireen Dubel and André Hielkema, 172–180. The Hague: HIVOS.

Verhoeff, Nann. 2012. *Mobile Screens: The Visual Regime of Navigation.* Amsterdam: Amsterdam University Press.

Visser, Robin. 2010. *Cities Surround the Countryside: Urban Aesthetics in Postsocialist China.* Durham: Duke University Press.

Wallerstein, Immanuel. 2006. *World-Systems Analysis: An Introduction.* Durham: Duke University Press.

Wan, Yanhai. 2001. "Becoming a Gay Activist in Contemporary China." *Journal of Homosexuality* 40, nos. 3–4: 47–64.

Wang Defeng. 2003. "'Xin Shanghai ren' yu dangdai Zhongguo de wenhua shengming" ['New Shanghai people' and the cultural life of contemporary China]. In *Xin Shanghairen* [New Shanghai people], edited by Shanghai Zhengda yanjiu suo. Hong Kong: Sanlian shudian.

Wang Guorong. 1998. *Kuashiji de Shanghai dushi wenhua xingxiang.* Shanghai: Shanghai shehui kexueyuan chubanshe.

Wang, Jianjun, Genya Yu, and Shuifeng Li. 2008. "The Implications of the Development of a Dam Economy in the United States for Nu River Hydropower Development: Dam Removals in the United States." In *Nujiang, Lancangjiang and Jinshajiang: Research on the Exploitation of Hydropower Resources and Environmental Protection*, edited by Yaohua He, 148–186. Kunming: Southwest Nationalities Research.

Wang, Jianming, and John A. Young. 2006. "Applied Anthropology in China." *National Association for Practicing Anthropology Bulletin* 25: 70–81.

Wang, Jing. 2010. *Brand New China: Advertising, Media, and Commercial Culture.* Cambridge: Harvard University Press.

Wang, Lan, Ratoola Kundu, and Xiangming Chen. 2010. "Building for What and Whom? New Town Development as Planned Suburbanization in China and India." In *Suburbanization in Global Society, Research in Urban Sociology, Volume 10,* edited by Mark Clapson and Ray Hutchison, 319–345. Bingley, UK: Emerald Group.

Wang, Shaoguang, and Angang Hu. 1999. *The Political Economy of Uneven Development: The Case of China.* Armonk, NY: M. E. Sharpe.

Wang Xiaoming, and Li Tuo, eds. 1999. *Zai xin yishixingtai de longzhao xia: Jiuling niandai de wenhua he wenxue fenxi.* Nanjing: Jiangsu remin chuban she.

Wang, Xuan, et al. 2012. "Urban Ecological Regulation Based on Information Entropy at the Town Scale: A Case Study on Tongzhou District, Beijing City." *Procedia Environmental Sciences* 13: 1155–1164.

Wang Ying. 2012. "Jiyu shengtai lunlixue jiaodu de Dujiangyan lüyou kechixu fazhan wenti" [Research on sustainable development of Dujiangyan tourism from the perspective of ecological ethics]. *Chengdu daxue xuebao* 4: 11–16.

Wei, Yehua Dennis, and Xinyue Ye. 2014. "Urbanization, Urban Land Expansion, and Environmental Change in China." *Stochastic Environmental Research and Risk Assessment* 28: 757–764.

Weller, Robert P., ed. 2005. *Civil Life, Globalization, and Political Change in Asia: Organizing between Family and State.* London: Routledge.

White, Sydney. 2010. "The Political Economy of Ethnicity in Yunnan's Lijiang Basin." *Asia Pacific Journal of Anthropology* 11, no. 2: 142–158.

Williams, Dee Mack. 1996. "Grassland Enclosures: Catalyst of Land Degradation in Inner Mongolia." *Human Organization* 55, no. 3: 307–313.

———. 2002. *Beyond Great Walls: Environment, Identity and Development on the Chinese Grasslands.* Stanford: Stanford University Press.

Wong, Andrew, and Qing Zhang. 2000. "The Linguistic Construction of the *Tongzhi* Community." *Journal of Linguistic Anthropology* 10, no. 2: 248–278.

World Commission on Dams (WCD). 2000. *Dams and Development: A New Framework for Decision-Making.* London: Earthscan.

Wu, Fulong. 2007. "Beyond Gradualism: China's Urban Revolution and Emerging Cities." In *China's Emerging Cities: The Making of New Urbanism,* edited by Fulong Wu, 3–25. London: Routledge.

Wu, Fulong, Fangzhu Zhang, and Chris Webster, eds. 2014. *Rural Migrants in Urban China: Enclaves and Transient Urbanism.* London: Routledge.

Wu Hung. 2005. *Remaking Beijing: Tiananmen Square and the Creation of a Political Space.* Chicago: University of Chicago Press.

Wu, Xiaogang, and Donald Treiman. 2007. "Inequality and Equality under Chinese Socialism: The Hukou System and Intergenerational Occupational Mobility." *American Journal of Sociology* 113: 415–445.

Wutich, Amber. 2011. "The Moral Economy of Water Reexamined: Reciprocity, Water Insecurity, and Urban Survival in Cochabamba, Bolivia." *Journal of Anthropological Research* 67, no. 1: 5–26.

Xiang Biao. 2010. "Putongren de 'guojia' lilun" [*Guojia*: A common people's theory of the state in contemporary China]. *Kaifang shidai* [Open Times], October: 117–132.

Xiaohe. 1999. *Huizi* [Golden boy]. www.bookbao.com/view/201102/01/id_XMTQ3MzYy.html.

Xiong Yuezhi. 2003. "Shanghai ren de guoqu, xianzai yu weilai" [The past, present, and future of Shanghai people]. In *Xin Shanghairen* [New Shanghai people], edited by Shanghai Zhengda yanjiu suo. Hong Kong: Sanlian shudian.

Xu, Jianchu, and Andreas Wilkes. 2004. "Biodiversity Impact Analysis in Northwest Yunnan, Southwest China." *Biodiversity and Conservation* 13: 959–983.

Yan, Hairong. 2008. *New Masters, New Servants: Migration, Development, and Women Workers in China*. Durham: Duke University Press.

Yan, Weiqiong, Daojie Wang, and Guojie Chen. 2011. "Reconstruction Strategies after the Wenchuan Earthquake in Sichuan, China." *Tourism Management* 32: 949–956.

Yan, Zhaoli, Ning Wu, Yeshi Dorji, and R. Jia. 2005. "A Review of Rangeland Privatization and Its Implications on the Tibetan Plateau, China." *Nomadic Peoples* 9, no. 1: 31–52.

Yang, Guobin. 2005. "Environmental NGOs and Institutional Dynamics in China." *China Quarterly* 181: 46–66.

Yang Guorong. 2003. "Shanghairen: lishi yu xiandai" [Shanghai people: history and the present]. In *Xin Shanghairen* [New Shanghai people], edited by Shanghai Zhengda yanjiu suo. Hong Kong: Sanlian shudian.

Yang, Mayfair Mei-hui. 1997. "Mass Media and Transnational Subjectivity in Shanghai: Notes on (Re)Cosmopolitanism in a Chinese Metropolis." In *Ungrounded Empires*, edited by Aihwa Ong and Donald Nonini.

———. 2008. *Chinese Religiosities: Afflictions of Modernity and State Formation*. Berkeley: University of California Press.

Yang Qingjuan, Li Bei, and Li Kui. 2011. "Rural Landscape Research in Chengdu's Urban-Rural Integration Development." 2011 International Conference on Green Buildings and Sustainable Cities. *Procedia Engineering* 21: 780–788.

Yao, Runfeng. 2004. "Shui Li Bu: Wo guo shuiku yimin zongshu 1,500 duo wan" [Ministry of Water Resources: China's dam-induced migrants total at least 15 million]. Beijing: *Xinhua News*.

Ye Xin. 1995. *Niezhai*. Television series. Shanghai: Shanghai dianshi tai.

———. 2001. "*Niezhai* huyu ban yinfa de zhenglun." Reprinted in Ye Xin, *Wo shengming de liangji*. Shanghai: Renmin chubanshe, 2004.

Yeh, Emily T. 2012. "Transnational Environmentalism and Entanglements of Sovereignty: The Tiger Campaign across the Himalayas." *Political Geography* 31: 418–428.

———. 2013. "Blazing Pelts and Burning Passions: Nationalism, Cultural Politics and Spectacular Decommodification in Tibet." *Journal of Asian Studies* 72, no. 2: 319–344.

Yeh, Emily T., and Gaerrang. 2011. "Tibetan Pastoralism in Neoliberalizing China: Continuity and Change in Gouli." *Area* 43, no. 2: 165–172.

Young, Jason. 2013. *China's Hukou System: Markets, Migrants and Institutional Change*. New York: Palgrave Macmillan.

Yu, Anthony, ed. and trans. 2012. *Journey to the West*, 4 vols. Chicago: University of Chicago Press, 2012.

Yu, Dan Smyer. 2012. *The Spread of Tibetan Buddhism in China: Charisma, Money and Enlightenment*. New York: Routledge.

Yu Jianrong. 2010. "Xinfang wenze de zhidu fansi" [Systemic reflections on petition-based accountability]. *Fanfang dushi bao* [The southern metropolis daily], March 1.

Yundannima. 2012. "From 'Retire Livestock, Restore Rangeland' to the Compensation for Ecological Services: State Interventions into Rangeland Ecosystems and Pastoralism in Tibet." PhD dissertation. University of Colorado, Boulder.

Zhan, Mei. 2009. *Other Worldly: Making Chinese Medicine through Transnational Frames*. Durham: Duke University Press.

Zhang, Li. 2001. *Strangers in the City: Reconfigurations of Space, Power and Social Networks within China's Floating Population*. Stanford: Stanford University Press.

Zhang, Li. 2010. *In Search of Paradise: Middle-Class Living in a Chinese Metropolis*. Ithaca, NY: Cornell University Press.

Zhang Rulun. 2003. "Shanghai ren de difang yishi he shijie yishi" [Shanghai people's regional consciousness and global consciousness]. In *Xin Shanghairen* [New Shanghai people], edited by Shanghai Zhengda yanjiu suo. Hong Kong: Sanlian shudian.

Zhang, Yongqing. 2011. "Special Issue: Online Literature in China: Present Situation and Theoretical Reflections." *Social Sciences in China* 32, no. 1: 182–190.

Zheng Jinran, and Yang Wanli. 2015. "Relocating to Tongzhou Seen as Wise." *China Daily*, June 24.

Zhou Yixing, and Laurence J. C. Ma. 2000. "Economic Restructuring and Suburbanization in China." *Urban Geography* 21, no. 3: 205–326.

Zhu Hong. 2001. "Yiju chengshi zhi qianshi jinsheng." *Shanghai wenhua* 51: 103–105.

Zhu, Jieming. 1999. *The Transition of China's Urban Development: From Plan-Controlled to Market-Led*. Westport, CT: Praeger.

Zhu Shengjian. 2001. "Shanghai ju, da buyi." *Shanghai wenhua* 51: 105–108.

Žižek, Slavoj. 1989. *The Sublime Object of Ideology*. New York: Verso.

CONTRIBUTORS

YOMI BRAESTER is Byron W. and Alice L. Lockwood Professor in the Humanities and Professor of Comparative Literature, Cinema and Media at the University of Washington in Seattle. He has published extensively on modern Chinese literature, film, and visual culture. Among his publications are the books *Witness against History: Literature, Film, and Public Discourse in Twentieth-Century China* and *Painting the City Red: Chinese Cinema and the Urban Contract*. His current book projects include *Archives of the Future: New Media and the Reinvention of Public Space* and *Cinephilia Besieged: Viewing Communities and the Ethics of the Image in the People's Republic of China*, which is supported by a Guggenheim fellowship.

ALEXANDER DES FORGES is Associate Professor of Chinese and Chair of the Department of Modern Languages, Literatures, and Cultures at the University of Massachusetts in Boston. He is the author of *Mediasphere Shanghai: The Aesthetics of Cultural Production*, and he has published several articles and book chapters on Chinese literature and film in comparative perspective. His current research focuses on literary economies in the early modern period, resulting most recently in "Sleights of Capital: Fantasies of Commensurability, Transparency, and a 'Cultural Bourgeoisie,'" *differences* 24, no. 3 (2014).

KABZUNG is an independent scholar. He grew up in a nomadic community on the southeastern Tibetan Plateau and received graduate training in both China and the United States. His research interests include development, religious movements, and environmental change in the pastoral societies of southeastern Tibet, China. He is the author of "Alternative (to) Development on the Tibetan Plateau: Preliminary Research on the Anti-Slaughter Movement" (*Revue d'Etudes Tibétaines*) and "Development as Entangled Knot: The Case of the Slaughter Renunciation Movement in Tibet, China" (forthcoming in the *Journal of Asian Studies*).

RACHEL LENG is Research Associate and Chief English Editorial Program Officer at the Asian Institute of Policy Studies in Seoul, South Korea. She graduated with a B.A. from Duke University and an M.A. in Regional Studies–East Asia from Harvard University. Her research interests focus on modern Chinese literature, society, and cultural studies. At Harvard, she was a coeditor-in-chief for the *Harvard Asia Quarterly*.

Her M.A. thesis on Sinophone production across Greater China and Southeast Asia won an Honorable Mention for the 2015 Joseph Fletcher Memorial Award. Recent publications include articles for university journals and "Eileen Chang's Feminine Chinese Modernity: Dysfunctional Marriages, Hysterical Women, and the Primordial Eugenic Threat" (*Quarterly Journal of Chinese Studies*, March 2014).

RALPH LITZINGER is Associate Professor of Cultural Anthropology at Duke University. Since 2007, he has directed the Duke Engage Migrant Education project, a research and civic engagement project located in Beijing. He is the author of *Other Chinas: The Yao and the Politics of National Belonging*, which explores the politics of ethnicity and marginality among ethnic intellectuals in the early Reform period. More recently, he has published on environmental politics in southwest China, rural-urban migration, and suicide as a form of protest in contemporary China. He is the editor of "Labor Question in China: Apple and Beyond," which appeared in *South Atlantic Quarterly* in 2013. He is the coeditor of "Self-Immolation as Protest in Tibet," a 2012 online issue of *Cultural Anthropology*. He is currently completing a book manuscript called *Migrant Futures: China from the Urban Fringe*.

LISA ROFEL is Professor of Anthropology and Director of the Center for Emerging Worlds at the University of California, Santa Cruz. She has written and coedited five books and numerous articles on China, including *Other Modernities: Gendered Yearnings in China after Socialism* and *Desiring China: Experiments in Neo-liberalism, Sexuality, and Public Culture*. She is completing a coauthored book titled *The Twenty-First-Century Silk Road: China and Italy in the Transnational Fashion Industry*; editing a book of translated essays by the renowned Chinese cultural studies scholar Dai Jinhua, titled *After the Post-Cold War*; and editing a collection of short stories by Cui Zi'en titled *Silver Bible of the W.C.*

CARLOS ROJAS is Professor of Chinese Cultural Studies; Gender, Sexuality, and Feminist Studies; and Arts of the Moving Image at Duke University. He is the author of *Homesickness: Culture, Contagion, and National Transformation*, *The Great Wall: A Cultural History*, and *The Naked Gaze: Reflections on Chinese Modernity*. He is the coeditor (with Eileen Cheng-yin Chow) of *The Oxford Handbook of Chinese Cinemas* and *Rethinking Chinese Popular Culture: Cannibalizations of the Canon*, and also the coeditor (with David Der-wei Wang) of *Writing Taiwan: A New Literary History*. He has translated book-length literary fiction by Yu Hua, Yan Lianke, and Ng Kim Chew.

BRYAN TILT is Associate Professor of Anthropology at Oregon State University. His research focuses on sustainable development, pollution control, and water resources in China and the United States. A former Fulbright Senior Research Scholar in Beijing, he is the author of the book *Dams and Development in China: The Moral Economy of Water and Power*, which examines China's push for hydropower development and its consequences for local communities. He is also the author of *The Struggle for Sustainability in Rural China: Environmental Values and Civil Society*.

ROBIN VISSER is Associate Professor of Chinese and Chair of the Department of Asian Studies at the University of North Carolina at Chapel Hill. She specializes in modern Chinese literary and cultural studies, urban studies, and environmental studies. Her book *Cities Surround the Countryside: Urban Aesthetics in Postsocialist China* analyzes Chinese urban planning, fiction, cinema, art, and cultural studies at the turn of the twenty-first century. Her current research is on Sinophone eco-literature.

BIAO XIANG teaches anthropology and migration studies at Oxford University. He is the author of *The Intermediary Trap*, *Global Bodyshopping* (winner of the 2008 Anthony Leeds Prize), *Transcending Boundaries*, and *Return: Nationalizing Transnational Mobility in Asia* (coedited with Brenda Yeoh and Mika Toyota). He has published about sixty articles, including one in *Pacific Affairs* that received the 2012 William L. Holland Prize for an outstanding article.

EMILY T. YEH is Professor and Department Chair of Geography at the University of Colorado in Boulder. She conducts research on nature-society relations in the Tibetan parts of the PRC, including projects on conflicts over access to natural resources, the relationship between ideologies of nature and nation, the political ecology of pastoral environment and development policies, the vulnerability of Tibetan herders to climate change, and emerging environmental subjectivities. Her book *Taming Tibet: Landscape Transformation and the Gift of Chinese Development* explores the intersection of political economy and cultural politics of development as a project of state territorialization. She is also the coeditor (with Chris Coggins) of *Mapping Shangrila: Contested Landscapes in the Sino-Tibetan Borderlands* and (with Kevin O'Brien and Ye Jingzhong) of *Rural Politics in Contemporary China*.

INDEX

Aba Prefecture, 110, 112
Abbas, Ackbar, 21
abjected figures, 153, 156–60
accumulation of worldly goods, 113–15, 117–18, 125–26
affective temporal engagements, 170, 171, 172, 177–84, 188, 189
agricultural practices, 92, 100–103
Ai Weiwei, 191, 203
Along the River during the Qingming Festival (*Qingming shanghe tu*), 27
alternative energy resources, 87, 89, 94, 106
American Chamber of Commerce, 57
animal husbandry, 109–29
anticipatory debt, 212
Anti, Michael, 223
anti-slaughter movement, 109–11, 115–29
architectural imagery, 23–24, 26–28, 30–34
Asian financial crisis, 5–6, 11
Augé, Marc, 20
autonomous designations, 91–92

Bao Zonghao, 68, 76–77, 78, 79
Barmé, Geremie, 2
ba sheng huo gaohao (decent life), 179
Baudrillardian critique, 19–20
Baudrillard, Jean, 19–20, 21
Beautiful New World (film), 62–63, 73, 82
Beijing: gay community, 152; Master Plan (2004–2020), 48–51; migrant schools, 192–93, 202; migrant workers, 186, 187, 206; reconstruction and gentrification, 15–16, 21–24, 26, 32–34; spatiotemporal overlaying, 16, 21, 24, 26, 30–34; suburbinization planning strategies, 39, 48–52; urban palimpsest metaphor, 15–17, 20–21, 24, 26, 30–34. *See also* Tongzhou New City
Beijing 2010 State of the Environment Report, 51
Beijing 2020 Master Plan, 48–51
Beijing Literature, 81
Beijing Municipality, 51–52, 53, 60
bendi (locals), 177, 182
Benjamin, Walter, 18
Berlant, Lauren, 10
biantai (abnormal citizens), 154
biaojuexin (determination), 146
biaotai (attitude), 146
Bijiang Dam, 94 (table)
billboards. *See* media walls
bingtai (diseased citizens), 154
Bingzhongluo Dam, 93, 94 (table)
Bird, Brad, 10–11
Bodies That Matter (Butler), 153
Boyer, Christine, 19
BoySky (*yangguang didai*), 162
BRICS nations (Brazil, Russia, India, China, and South Africa), 11
Brodsky, Joseph, 26
bsod nams (principle of fortune), 110, 127

Buddhism: economic development alternatives, 123–29; modernist movements, 111, 116, 118–19, 128–29; oaths and vows, 120–23, 130; principles of karma, 110, 118, 120, 125, 127–28; slaughter renunciation movement, 109–11, 113, 115–29
building construction, 191–92
Bureau of International Economy, 140, 142, 145, 146
Bureau of Public Security, 58, 140, 141, 145
Butler, Judith, 153, 157

carbon emissions, 60, 88, 94
Caruth, Cathy, 209
case studies: affective temporal engagements, 177–84; fake visa and recruitment incident, 135–36, 138–47; origin stories, 174–77; transnational encounters, 184–87
causal model of reference, 216–17, 223
CCTV Tower, 21
cehua (strategic planning), 40
Celestial Jerusalem versus Earthly Jerusalem, 34
Cell Phone (novel), 215
cell phone use, 215, 223
Central Business District (Beijing), 22, 49
Certeau, Michel de, 18–19, 26
chaiqian (demolishing and relocating), 107
changgui (factory rules), 178
Chan, Jenny, 169, 180
chao di (stir-fry the land), 107
chengbao tian (contract land), 101
Chengdu, 112
Chengdu Municipality, 53, 56, 57, 60
Chengdu Municipal Planning Department, 53
chenggong renshi (successful individual), 73, 82
chengjiu gan (accomplishment), 177

Cheng Naishan, 63
Chengshi ye shi women de (novel). See *City Is Also Ours, The* (novel)
chengxiang tongchou (urban-rural integration), 37
Chen Ran, 151
Chen Sihe, 64, 68, 69, 71, 72, 73, 75–76, 77, 78
Chen Xueming, 66
Chien Shiuh-Shen, 42
chiku (eat bitterness), 186
children's rights, 193–94
China Blue (film), 211
China Daily, 49, 104
China Huadian Corporation, 93
China Real Estate Post, 42
China Southern Power Grid Company, 107
Chinese City Science Association, 42
Chinese Communist Party (CCP), 131, 182–83
Chinese-language websites, 151–52, 162, 214–15
Chongfu, 184
Chongqing, 53, 58, 81
chongxieben. See palimpsest metaphor
Choujiuo dongchang (novel). See *Enter the Clown* (novel)
chronotopes, 16–17
Chu, Julie, 183
Chu T'ien-wen, 151
cinematic panoramas and openings, 26–28, 30
City Is Also Ours, The (novel), 221
Clark, Marco, 92
class distinctions, 68, 73–74
Classification and Diagnostic Criteria of Mental Diseases, 154
clean earnings, 126–27
clean energy, 94, 104–5
closed-society myth, 171
coal industry, 93–94, 105
Coal Metro Travel Agency, 138, 142

248 INDEX

Comaroff, Jean, 128
Comaroff, John, 128
communications technologies, 215
Communist Manifesto, The, 1
Communist Party, 131, 182–83
compensation policies, 98–99, 101–2, 106, 107, 133
compulsive reenactments, 208–9
compulsory education, 124–25
consumerism, 113–15, 117–18, 125–26, 186
contract land, 101
Crystal Boys (novel), 150, 151
Cui Zi'en, 151
cult of personality, 2–3
Cultural Revolution (1966–1976), 37
culture of disappearance, 21

dagong (manual labor), 175
daibiao (representative), 77
Dai Shenzhi, 53, 57
Dai Zhikang, 65, 71, 75, 81, 82
Dalai Lama, 114
Dalian, 40
dams: benefits, 104–5; design specifications, 94 (table); economic considerations, 95–96, 98–99, 101–2, 106; environmental impacts, 87–89; feasibility reviews, 103–4; household surveys, 97–99; human displacement and resettlement, 88, 89, 93, 96–102; market-driven strategies, 100; moral considerations, 89, 103–6; state agencies, 92–93, 100
Dandelion School, 192–93, 202, 203
Daxing International Airport, 202
Dayanzhen, 59
day-trippers, 56
debt, 209, 212
deceased school children, 191–92
deictic, 220–21
demolition-and-relocation projects. See Qianmen Avenue reconstruction project

Deng Weizhi, 68
Deng Xiaoping, 1–2, 7, 206, 220
Derrida, Jacques, 3–6
descriptive model of reference, 216–17, 223
de Soto, Hernando, 103
developmentalism paradigm, 131
Develop the West strategy, 95
dialect speech, 65–66
Diaoyu/Senkaku Island disputes, 132
digital archives, 213–15
digital cameras, 30–31
dingzi hu (nail households), 74
Dishui Lake, 44, 45, 46
dispossession, 103
ditan jingji (low-carbon economy), 87–88
ditan nengyuan (low-carbon energy), 94
documentary films, 33, 144, 149, 194–203, 211
Doje Cering, 40
domestic clothing industry, 172, 178, 181, 184–86
dormitory labor regime, 180
Duan Jin, 40
Dujiangyan, 53
Dujiangyan Ecological Garden City, 55
Dujiangyan New City, 39, 55–57, 60, 61
Duke Engage project, 203
Duplicated Memories (2008), 15

earthquakes, 42, 53, 104, 108, 191–92
East China Sea Bridge, 44
eco-city development: Dujiangyan New City, 39, 53, 55–57, 60, 61; Lingang New City, 39, 43–48, 60, 61; province-manage-county (PMC) reform, 58; suburbinization planning strategies, 39–40, 41 (table), 42–43; Tongzhou New City, 39, 48–52, 60, 61
ecologically sustainable techno-utopias. *See* Lingang New City
Ecosystem Services Value (ESV), 56
educational opportunities, 192–204

INDEX 249

Eiffel Tower, 34
ejaculate, 158–59
electric-power sector, 93, 100, 104–5
electronics industry, 169
Eleventh Five-Year Plan (2006–2011), 58, 95, 104
elite entrepreneurs, 93, 170–71, 175–77, 179, 186–89
Ellingson, Stephen, 135
energy consumption, 87–88, 93–94
Engels, Friedrich, 1
Enter the Clown (novel), 151
entrepreneurship: ethnic minorities, 126–29; land development projects, 40, 42–43; migrant workers, 186, 189; Shanghai, 76–81, 186; textile and garment industry, 170, 171, 182–83, 186
Environmental Impact Assessment Law (2004), 91, 98, 103, 105
environmental nongovernmental organizations (NGOs), 91
ephemeral transcendence, 183
ethnic minorities: entrepreneurship, 126–29; hydropower development projects, 89, 97; Marxist classifications, 91; Nujiang Prefecture, 92, 96; urban-rural integration plans, 59
ethnography of incidents, 134–36
European debt crisis, 11–12
evil cults, 114
exhortations, 78
expropriation, 103
extralegal land use, 102–3

fake visa and recruitment incident, 135–36, 138–47
family estrangement, 197–99, 211–12
Fan Lixin, 196, 197
farming households, 100–103
fashion clothing import-export business, 168–88
Fengxian New District, 46

Feng Xiaogang, 215
figures of abjection, 153, 156–60
first-generation new cities, 41 (table)
Fish Bowl/Tiananmen (1999), 15
flash drive as metaphor, 212–15, 217–19
folk theory of the state, 148
forced relocations, 55, 56–57, 59
Ford, Jocelyn, 194–95
foreign direct investment, 168
foreign exchange reserves, 7
fossil fuel use, 87–88, 93–94, 105, 106
Frege, Gottlob, 216
Freud, Sigmund, 18, 208–9
Friedman, Eli, 193
Fugong Dam, 94 (table)
fupin wuqi (poverty alleviation weapon), 96
fupin zhanlüe (poverty alleviation strategy), 96

gaige kaifang. *See* Reform and Opening Up campaign
gangtai (Hong Kong/Taiwan) music and film, 63
Gansu Province, 205
Gaoligong mountain range, 88–89
Gao Xiaowei, 49
Garden City model, 43, 53
gay community, 150–62, 163
gendered industries, 177–78
gender issues. *See* LGBT community
Ghana, 128
Gini coefficient, 8
global capitalism, 174
global urbanism, 36–37
Golden Projects, 214
Golden Shield Project, 214
Golog Prefecture, 123
government corruption, 136
Grace, Helen, 30–31
Graeber, David, 90
grassland use rights contract system, 111–12

Great Britain, 220
Great Firewall of China, 214
Great Leap Forward (1958–1961), 219–20
Great Shanghai Plan (1946), 43
green city development. *See* eco-city development
green energy, 94
Green Metro Business Development Consulting Center, 138, 142, 145
gross domestic product (GDP), 7
Guangdong Province, 180, 182
Guangpo Dam, 93, 94 (table)
Guangzhou, 196–97, 211
Guangzhou New Town, 40
guanxi networks, 79
guihua (urban planning), 40
Guitu lieche (documentary). *See Last Train Home* (documentary)
guojia (the state), 132, 144–45, 147, 148
Gu Xiaoming, 73

Habermas, Jürgen, 207
Haining, 184
Hanchao Lu, 75
Han Chinese, 59, 91, 109, 113, 126
Hangzhou, 42, 173, 177, 180, 186
Hangzhou Silk Corporation, 173
Hankiss, Elemér, 80
Harvey, David, 17, 19, 99, 174
Hebei Province, 200
Hegel, Georg Wilhelm Friedrich, 82
Hegelian influence, 64, 72, 78, 188
Henan Province, 205
Hershatter, Gail, 176
heterosexuality, 159–61
He Zhenzong, 221
high-voltage direct-current (HVDC) transmission lines, 96
historical imagination of overcoming, 183
HIV/AIDS pandemic, 151
home-based industries, 177
homoerotic literature, 150–62, 163
homophobia, 152

homosexuality, 150–62, 163
Hong Kong, 63, 67
Hong Kong Gay and Lesbian Film Festival (*xianggang diyi jie tongzhi dianying jie*), 163
Hong Kong Retrocession, 5
Hongyuan County, 112–13
Hongyuan County, Aba Prefecture, Sichuan Province, 110
hooliganism, 153–58
hopefulness, 147
Household Responsibility System, 100
household surveys, 97–99
Housing Project for Herders, 114–15, 124–25
Howard, Ebenezer, 43
Hsing You-tien, 40
Huadian Corporation, 93
Hua Min, 71
Huang, Michelle, 66, 68
Huangren shouji (novel). *See Notes of a Desolate Man* (novel)
Hui Chinese, 109, 113, 126
Huizi (Internet novella), 152–53, 155–61, 162, 163
Hu Jintao, 59, 131, 187
hukou (household registration) system: Beijing, 199; characteristics and functional role, 2, 176, 206–7, 214; children's rights, 193; educational opportunities, 192; rural land development policy, 58, 59, 107, 176, 188; rural-to-urban migrants, 9, 206–8, 215; Shanghai, 74; social identity, 217–18
Human Development Index (HDI), 95
humanist spirit debates, 63–64
Hung Ho-fung, 180
Hu Shouyun, 71
Huyssen, Andreas, 19
Huzhou, 175, 176
hybrid communist-capitalist economic system, 1–3, 5–8, 10, 167–73, 188–89

hydropower use and development: benefits, 104–5; design specifications, 94 (table); economic considerations, 95–96, 98–99, 101–2, 106; environmental impacts, 87–89; feasibility reviews, 103–4; household surveys, 97–99; human displacement and resettlement, 88, 89, 93, 96–102; market-driven strategies, 100; moral considerations, 89, 103–6; Southwest China, 87–89; state agencies, 92–93, 100
hyperurbanization, 21
hypervisual space, 32–34

I Am Liu Yuejin (novel), 205–23
imagined futures, 170, 171, 181–84
Impression Liu Sanjie (*Yinxiang Liu sanjie*), 56
incidents, ethnography of, 134–36
independent schools, 192–93
indigenous populations, 92; . *See also* ethnic minorities
individual identity, 216–18
Indonesia, 128
informal land use, 102–3
information management systems, 213–15
ink scroll display, 24, 25, 26–27
in-migration impacts, 62–66, 68–70, 73–77, 80, 82
Inner Mongolia, 42
instrumentalism paradigm, 133–34
interactive writing, 162
International Commission on Large Dams, 87
international labor recruitment, 137
International Monetary Fund (IMF), 5–6, 11–12
International Rivers, 108
Internet literature, 151–52, 161–62, 163
Internet monitoring, 214–15
IOU as metaphor, 209–10, 213–14, 215, 217

irrigation systems, 53, 61
Islam, 128
Italian fashion clothing factories, 168, 169, 172–73, 178

Jacobson, Roman, 220
Jiangsu Province, 182
Jiang Zemin, 83, 131
jiaoqu (suburb), 37
Jiaxing, 173, 176, 180, 181, 186
Jia Zhangke, 34
Ji Dan, 199–202
Jigme Phuntsok, 110, 115, 116–17, 118, 119, 129
jihua (economic planning), 40
Jinhua, 174–75, 176
Journey to the West (novel), 222
jumbotrons, 22, 31
jumin hukou (residential house hold registration), 107
Jung In Kim, 45, 47

Kangxi Emperor Returning to the Forbidden City from His Southern Expedition, The, 26
karma, 110, 118, 120, 125, 127–28
Khenpo Jigme Phuntsok, 110, 115, 116–17, 118, 119, 129
khenpos, 111, 115–16, 118–20, 125–27, 129–30
Khenpo Tsultrim Lodroe, 110, 117, 118, 119–20, 122–26
Kim Jung In, 45, 47
Kojève, Alexander, 4–5, 6
Kong Shuyu, 81
Koolhaas, Rem, 21
Kripke, Saul, 216
Kristeva, Julia, 158–59, 160

labor out-migration, 136–38
labor protests, 133, 180–81
Lacan, Jacques, 21, 155, 156
lamas, 129

Lancang River, 107
Land Administration Law, 38
land use, 102–3
language dialects, 65–66
laojia (family home), 182
large-scale dam construction. *See* hydropower use and development
large-scale video, 22–23, 27, 30–34
Larung Gar. *See* Serthar Buddhist Institute (Serthar Gar)
Last Train Home (documentary), 196–99, 211
law of cause and effect, 110, 118, 120, 125, 127–28
Lawrence, Regina, 135
LED screens, 22–23, 27, 30–34
Lee, Ching Kwan, 133, 180, 182
Lefebvre, Henri, 36–37
lesbian community, 151, 154. *See also* LGBT community
LGBT community, 150–62, 163
Liaoning Province, 135, 145
lilong neighborhood, 65
Lin Bai, 151
Lingang New City, 39, 43–48, 60, 61
Lingang New District, 46
Ling Yan, 151
Litzinger, Ralph, 69, 168–69, 179, 211
Liu Hongkuan, 24, 26–27
Liuku Dam, 93, 94 (table), 102, 104
liumang (hooligan), 153–54
liumangzui (hooligan law), 154
Liu Qibao, 124
Liu Zhenyun, 206–7, 215, 219–21
livestock industry, 109–29
local-state government contradictions: characteristics, 133–34; fake visa and recruitment incident, 135–36, 138–47; hopeful engagement, 147
local worldview, 71–73, 77–80
low-carbon cities, 42, 51
low-carbon economy, 87–88
low-carbon energy, 94

low-income housing projects, 52, 55
Lu Hao, 15
Lumadeng Dam, 94 (table), 97
Lunyu, The (The analects), 207
luohou (backward), 91
Luo Tingting, 49
lüse nengyuan (green energy), 94
Lushui County, 98
Lushui Dam, 94 (table), 97

MacGuffin, 212, 223
Maji Dam, 93, 94 (table), 97, 104
Manovich, Lev, 33
Manwan Dam, 107
Maoist agricultural system, 100
Mao Zedong, 1–3, 7, 219–20
marine eco-cities, 42. *See also* Lingang New City
Maritime Museum, 47
market economy: agricultural practices, 92, 97; electric-power sector, 99–100; free-market principles, 2, 3, 168; functional role, 90; *guanxi* networks, 79; hydropower development projects, 89; neoliberalism, 99, 114, 127; out-migration, 136–38; property rights, 100, 102–3, 114; secular market, 114, 115; Shanghai, 64, 67, 69–73; social life, 132; spatiotemporal fragmentation, 17; yak-based economic development, 111–13, 124–28
Marketplace (radio program), 195
Marx, Karl, 1
mass incidents, 135
materialistic standard of living, 113–15, 117–18, 125–26, 171
media walls, 22–23, 27, 30–34
Meili xin shijie (Beautiful new world). See *Beautiful New World* (film)
Meishi Street (documentary), 33
Mekong River, 95, 107
Meys, Olivier, 33
migrant education, 192–204

INDEX 253

migrant labor: characteristics, 9–10; educational opportunities, 192–204; family estrangement, 197–99, 211–12; identity references, 218; in-migration impacts, 62–66, 68–70, 73–77, 80, 82; out-migration, 136–38; rural-to-urban migrants, 9–10, 59, 168–89, 192–94, 206–7; Shanghai, 68–69, 73–74, 76, 79, 187; socioeconomic factors, 206–8, 218; textile and garment industry, 168–89; urban-rural integration plans, 59
mingong (migrant workers), 68–69, 73–74, 76, 79
Ministry of Commerce, 144
Ministry of Construction, 37
Ministry of Electrical Power, 107
Ministry of Housing and Urban-Rural Development, 37
Ministry of Public Security, 144
Ministry of Water Resources, 88, 93, 104
minjian jiyi (popular cultural memory), 76
minjian qiye (nonstate enterprises), 79
minjian yanjiu jigou (nonstate research organization), 80–82
minority nationalities. *See* ethnic minorities
Min River, 53
minying qiye jia (private entrepreneurs), 76–80
minzhong (masses), 67, 68, 78
mirror-stage theory, 155, 156
Mission: Impossible—Ghost Protocol (film), 10–11
mixin (superstition), 114
Miyazaki, Hirokazu, 136
mobile devices, 28, 30–31
monasteries, 129
monetary debt, 209, 212
monks, 129
Moore, Sally, 134

moral economy: basic principles, 90; hydropower development projects, 89, 103–6
moralism paradigm, 133–34
Motorola, 215
multinational corporations, 168–69
Myanmar (Burma), 88, 92, 107

Naftali, Orna, 193
naming, theories of, 216–17, 219–21
Nanhui New City, 46
Nanjing, 42
Nanyuan Airport, 192, 202
National Comprehensive CURD (Coordinated Urban-Rural Development) Experimental Zones, 53, 58
National Development and Reform Commission, 93, 103, 104
nationalism paradigm, 131–32
neoliberalism: commodity-chain production, 180; cultivation of desires, 114, 170; eco-city development, 42, 60; governance strategies, 99; secular neoliberalism, 111–23, 127–29; socioeconomic factors, 208; socioeconomic processes, 3, 6, 10; Tibetan yak herders, 109, 111–29; urbanization, 14, 16, 36
networked databases, 214
new city/new town development. *See* eco-city development
New Shanghai People, 62–82
New Urbanism, 17–18
New Xiaoshaba Village, 102
Niezhai (*Sinful Debt*). *See Sinful Debt* (television series)
Niezi (*Crystal Boys*). *See Crystal Boys* (novel)
9+3 Vocational Program, 125
Nine Year Compulsory Education Program, 125
nongmin (peasants), 181
nongovernmental organizations (NGOs), 91, 99, 103

non-places, 20
Nora, Pierre, 19
Northeast China, 135–36
Notes of a Desolate Man (novel), 151
Nowhere to Call Home (documentary), 194–96
Nujiang Prefecture, 92, 96, 97
Nuozhadu Dam, 107
Nu River, 88–89, 93
Nu River Project, 93, 96, 97–98, 101, 103–5
Nu River Valley, 90, 92, 97–99, 101
nuxing (slave quality), 66
nye ba'i rgyu (proximate causes), 128
Nyingma school of Tibetan Buddhism, 109, 115, 129

oaths and vows, 120–23, 130
OCT Contemporary Art Terminal, 15
'od dpog med kyi rgyud bzhi bo'i. See Sukhavati paradise of the Amitabha Buddha
Office for Letters and Visits, 140, 143, 144, 145
Official Penal Code (1957), 154
Old Shanghai identity, 65–66, 69, 74
Olympic Games (2008), 22, 23, 49, 55, 198, 203
Ong Aihwa, 99, 114
Open Up the West campaign, 109, 112
Ordos, 21, 42
origin stories, 169–70, 172, 174–77, 188, 189
Ou Ning. See Meishi Street (documentary)
out-migration, 136–38; . *See also* fake visa and recruitment incident
overgrazing, 111–12
overtime work, 180–81, 185

Pai Hsien-yung, 150, 151
pailou (five-gate memorial arch), 24
painted barriers, 27, 28, 29

Palatial Capital, The (*Tianqu danque*), 24, 25, 26–27
palimpsest metaphor, 15–21, 24, 26, 30–34
Passport Law (2007), 136
pastoralism. *See* yak-based economic development
Pearl River Delta, 63, 64, 81, 96
Peled, Micha, 211
Pentecostalism, 128
People's Daily, 135
Petition (documentary), 144, 149
petition system, 133, 144–45, 149
phonescapes, 30–31
Plexiglas installations, 15
politically based economic solutions, 133
political subjectivity, 148
politicization of abjection, 157
politics of emergence, 21–23, 26, 32–34
postquake building inspections, 192
postquake reconstruction, 39, 53, 55–57
poverty alleviation strategies, 96, 97
power generation companies, 93, 100, 104–5
prejudicial legacies, 153
prescriptive chronotopes, 17
principles of karma, 110, 118, 120, 125, 127–28
private entrepreneurs, 76–80
Private Life, A (novel), 151
private schools, 192–93
privatization programs, 1, 30, 93, 111, 114, 168, 173
pronoun use, 220–21
proper names, 216–17, 219–21
property rights, 56, 100–103, 106
province-manage-county (PMC) reform, 58
public schools, 192–93
Pudong New District, 44, 63
Pun Ngai, 169, 180, 186

Qianmen Avenue reconstruction project, 23–24, 26–28, 30–34
Qianmen East Road, 24
Qianmen Hotel No. 1 (video artwork), 33
Qian Men Qian: A Disappearance Foretold (documentary), 33
Qiao Yu, 81
qibu (Seven Don'ts), 78
Qinghai Lake, 123, 127
Qinghai-Tibet Plateau, 88–89
qingjie nengyuan (clean energy), 94
Qin's story, 196–99
Qiu Miaojin, 151
quasi-owned housing, 50
queer community, 150–62, 163

"race to the bottom" production strategy, 168
Railton, Jeremy, 22
Rakhor Monastery, 119
Rakhor Village, Sichuan Province, 110, 115, 119–20, 124
real estate development. *See* eco-city development
rectification of names, 207
Reform and Opening Up campaign, 7, 90–93, 101, 111, 151, 206, 220
Reform Era: *guanxi* networks, 79; hydropower development projects, 92, 99–100; income inequalities, 7–8, 96, 177; in-migration impacts, 70
regime stability, 131–32
relay writing, 162
religious beliefs and practices, 114, 128. *See also* Buddhism
religious elites, 110, 119, 127, 129
renwen jingshen (humanist spirit) debates, 63–64
reservoir-induced seismicity, 108
responsibility land, 101
returned youth, 63, 64, 69–70
rightful resistance paradigm, 133
ring ba'i rgyu (remote causes), 128

Rofel, Lisa, 74, 114
Rojas, Carlos, 69
Rose, Nikolas, 113
Roy, Ananya, 37–38
Ruan Yisan Foundation for the Protection of City Heritage, 83
rural land conversion, 38–40, 48, 53, 57, 58–60. *See also* eco-city development
rural land property rights, 100–103, 106
rural-to-urban migrants, 9–10, 59, 168–89, 192–94, 206–7
Russell, Bertrand, 216

Sahlins, Marshall, 134
Saige Dam, 94 (table), 104
Salween River, 88, 107
same-sex relationships, 150–62, 163
san nian ziran zainan (three years of natural disaster), 219
satellite towns. *See* eco-city development
school children deaths, 191–92
school construction, 191–92
screen walls, 22–23, 27, 30–34
scroll display, 24, 25, 26–27
sdig pa (killing sin), 109
second-generation new cities, 41 (table)
secular capitalist development, 109–11, 113–15, 126–29
secular neoliberalism, 111–23, 127–29
self-abjection, 158–60
self-consciousness, 80
semen, 158–59
Send Western Electricity East plan, 96
sentient beings, 113, 116–18
Serthar Buddhist Institute (Serthar Gar), 115–16
Serthar County, Sichuan Province, 110, 115
Sewell, William, 135
sexual identity, 150–53, 157–58
sexual openness, 152, 154

Shangfang (documentary). See *Petition* (documentary)
Shanghai: city-nation contrasts, 64–70; class distinctions, 68, 73–74; dialect speech, 65–66; economic and cultural development and identity, 63–82; entrepreneurial efforts, 76–81, 186; gay community, 152; in-migration impacts, 62–66, 68–70, 73–77, 80; local worldview, 71–73, 77–80; migrant workers, 68–69, 73–74, 76, 79, 187; real estate development, 74–75, 81–82; suburbinization planning strategies, 39, 43–48; textile and garment industry, 171, 173; Western influence, 66–67. *See also* Lingang New City
Shanghai Financial News, 46
Shanghai Municipality, 39, 42, 44, 46, 60
Shanghai Ruan Yisan Urban Planning and Design Company, Limited, 83
Shanghai shi Zhongguo de Shanghai (Shanghai is China's Shanghai), 67
Shanghai Zhengda Investment Group, 75
Shanghai Zhengda Research Center, 75, 80–82
Shanxi Province, 205
shaoshu minzu (minority nationalities), 91
Shenbei, 42
Shenyang Labor Center, 139, 141, 142
Shenyang Province, 42
Shenzhen, 15
Shibuya (Tokyo), 22
shift system, 179–80
shifu (master craftsperson), 175, 178, 184
shijie yishi (world-consciousness), 77
Shi Lishan, 104
Shisheng huamei (novel). See *Silent Thrush, The*
Shitouzai Dam, 94 (table)
Shouji (novel). See *Cell Phone* (novel)
shua liumang (plays with hooliganism), 156

Shue, Vivienne, 131
shuidian jidi (hydropower base), 95
shunying, 70
Sichuan Province, 53, 110–29, 181, 191–92. *See also* Dujiangyan New City
Siegel, Greg, 33–34
Silent Thrush, The (novel), 151
Silk Road, twenty-first-century, 172–73
silk-weaving factories, 173, 177, 180, 183
Simone Giostra & Partners, 22
simulacrum, 20, 34
sinful activities, 116–17, 119–20, 122–23
Sinful Debt (television series), 63, 65, 69
Singapore, 67
Sinohydro Corporation, 107
Sino-Swedish Environmental Technology Cooperation, 42
Siren shenghuo (novel). See *A Private Life* (novel)
skyscreens, 22–23, 27, 30–34
slaughter renunciation movement, 109–11, 113, 115–29
smartphones, 30–31
Smith, Nick, 38, 39, 58
social fluidity, 218
social identity, 207, 216–18
social inequalities, 167–68, 170–71, 177, 186–89
socialism from afar, 114
socialist-capitalist economic dichotomy, 1–3, 5–8, 10, 167–73, 188–89
socialist urban modernity, 59
social norms, 90, 153, 157; . See also LGBT community
Social Order Statute, 154
social production, 193
social tensions, 131–32
socioeconomic factors, 206–8, 218–20
Soja, Edward, 20
Solinger, Dorothy, 170–71
sonam (principle of fortune), 110, 127
Songjiang New City, 48
Songta Dam, 93, 94 (table), 104

INDEX 257

South Atlantic Quarterly (*SAQ*), 168, 179
Southwest China. *See* hydropower use and development
spiritualism, 4
Stalinist-style architecture, 59
State Compensation Law, 141
State Council, 101, 104
state education programs, 124–25
State Electric Power Corporation, 93, 100, 107
State Environmental Protection Administration, 103–4
state-managed religion, 114
structural chasm: basic concepts, 131–34; ethnography of incidents, 134–36; fake visa and recruitment incident, 135–36, 138–47; out-migration marketization, 136–38
suburbinization planning strategies, 37–40, 41 (table), 42–43, 48–52, 59–60
Sukhavati paradise of the Amitabha Buddha, 120
Sun Wukong, 222
Sun Yat-sen, 154
superstition, 114
Supertron screens, 22
Sustainable Cities, 42
sustainable development. *See* eco-city development
suzhi (quality), 58, 59, 124–25
Sweden, 42
swidden agriculture, 92

Taipei, 150
Taiping Rebellion, 4
Taiwan, 63, 67, 150–51
Tang Qing's story, 181–82
Tan Tan, 33
Tan Xuewen, 43
Tasang Dam, 107
temporal-spatial migration, 170–72, 174–89

Ten Years' Action on Education Construction Program, 125
textile and garment industry, 168–88
Thailand, 92
Thames Town, 48, 60
third-generation new cities, 41 (table)
Thompson, E. P., 90
Three Gorges Dam, 44, 88
three years of natural disaster, 219
Tiananmen Gate, 15, 26
Tiananmen Square, 22, 23, 31
Tianjin New Eco-City, 40, 42
tianming (heavenly mandate), 147
Tibet: economic development alternatives, 123–29; government control, 91–92; hydropower development projects, 93, 94 (table), 104; slaughter renunciation movement, 109–11, 113, 115–29; state education programs, 124–25; yak-based economic development, 109–29
Tibetan Plateau, 109
Times Square (New York), 22
tofu-dregs construction, 191
Tongji University, 53, 57, 59
Tongxiang, 173, 174, 176, 177, 179, 184, 186
tongxinglian (homosexual), 163
tongyi duo minzu guojia (unified multi-ethnic state), 91
tongzhi (gay) community, 152, 154, 156–62, 163. *See also* LGBT community
Tongzhou 2012 Master Plan, 51
Tongzhou New City, 39, 48–52, 60, 61
tourism, 26, 45, 55–56, 57, 112
tragedy of the commons, 111
transit-oriented development (TOD) satellite towns, 38, 39
transnational encounters, 171–72, 184–87, 188–89
transnational supply-chain factories, 168–89
transportation networks: Dujiangyan New City, 56; hydropower develop-

ment projects, 95; Old City (Beijing), 15; Shanghai, 48; Tongzhou New City, 49, 51; utopian cities, 60; yak-based economic development, 112, 113
traumatic neurosis, 208–9
tshe thar, 117
tshogs bsags bsod nams, 128
Tsing hua-MIT China Planning Network conference, 53
Tsinghua University, 59
Tsultrim Lodroe, 110, 117, 118, 119–20, 122–26
Turner, Victor, 134
Tweedie, James, 23
Twelfth Five-Year Plan (2011–2015), 42, 95, 104

UN Convention on Biodiversity, 92
UNESCO World Heritage Sites, 53, 61
United Nations Development Programme (UNDP), 95–96
urban consumption, 186
urban elites, 171, 215
urbanization: Baudrillardian critique, 19–20; eco-city development, 39–60; Lefebvrean perspective, 36–38, 43; palimpsest metaphor, 15–21, 24, 26, 30–34; politics of emergence, 21–23, 26, 32–34; population growth, 8, 9–10; spatiotemporal perspectives, 18–20; suburbinization, 37–38; urban-rural integration plans, 58–60
Urban Real Estate Development and Management Law, 38
Urban-Rural Planning Law (2007), 37
user-generated cityscapes, 30–31

value systems, 90
vegetarianism, 116, 130
Virilio, Paul, 21, 23
virtual urban-rural environments, 59–60

Visser, Robin, 191
Von Gerkan Marg und Partner, 44

waidi ren (outsiders), 69, 182
Wallerstein, Immanuel, 174
wall screens, 22–23, 27, 30–34
Wang Anyi, 63
Wang Defeng, 68, 70, 71, 80
Wang Guorong, 63
Wanjiajing, 183
wan (to play), 175
War of One's Own, A (novel), 151
water resources, 90
Wenchuan earthquake, 42, 53, 54, 191–92
wenhua gainian (cultural concept), 80
wenhua zaiti (vehicles of culture), 73
Wen Jiabao, 103, 104
Wenzhou, 176, 177, 178
West Sichuan Plain, 53
When the Bough Breaks (documentary), 199–202
wholly foreign-owned enterprises (WFOES), 173
Wo chao (documentary). See *When the Bough Breaks* (documentary)
Wo jiao Liu Yuejin (novel). See *I am Liu Yuejin* (novel)
workforce instabilities, 181
working conditions, 179–81, 185, 187
workplace transgressions, 180
World Bank, 87, 105
World Commission on Dams (WCD), 87, 106
world-consciousness, 77–80
World Conservation Union, 87
"worlding" projects, 171, 188, 189
World Summit of Sustainable Development (2002), 42
World, The (2004), 34
World Trade Organization (WTO), 173
wu da fadian jutou (electricity companies), 93

Wu Fulong, 48
Wujiaochang Wanda Plaza (Shanghai), 22

xianggang diyi jie tongzhi dianying jie: . See Hong Kong Gay and Lesbian Film Festival (*xianggang diyi jie tongzhi dianying jie*)
xiao chanquan fang (quasi-owned housing), 50
Xiaohe, 152
Xibu dakaifa (Develop the West strategy), 95
Xicui Entertainment complex, 22
xi dian dong song (Send Western Electricity East) plan, 96
xiejiao (evil cults), 114
xin buping (unhappiness), 178
xincheng (rural land conversion), 39–40
xing kaifang (sexual openness), 154
Xinhua News Agency, 88
Xin Shanghairen (New Shanghai People). See New Shanghai People
Xiong Yuezhi, 63, 68
Xiyouji (novel). See *Journey to the West* (novel)
xuetou (cave head), 79
xungen (roots-seeking) movement, 63

Yabiluo Dam, 94 (table), 97, 104
yak-based economic development, 109–29
Yanagisako, Sylvia, 172
yangguang didai (BoySky website), 162
Yang Guorong, 72
Yang, Mayfair, 67, 79, 80
Yang Qingjuan, 56
Yang Qing's story, 194–96

Yangshan Deepwater Port, 44, 45
Yangtze River, 95
Yangtze River Delta, 67, 181, 182
yanguang (global vision), 72
Yang Xifeng, 44–45
Yan Hairong, 169, 171, 183, 186, 187
Yansangshu Dam, 94 (table)
yarn-spinning factories, 173, 180, 181
yearnings, 171
Yellow River, 95
Yigeren de zhanzheng (novel). See *War of One's Own, A* (novel)
Yin Ming, 40
Yu Jianrong, 133
Yunnan Province, 59, 88, 92, 95–96

zeren tian (responsibility land), 101
Zhan, Mei, 171
Zhang Li, 114
Zhang Rulun, 64, 67, 71, 72, 73, 77–78
Zhang Yaxuan, 33
Zhang Yimou, 56
Zhao Liang, 144
Zhejiang Province, 173–87
Zhenfu silk-weaving factory, 173, 180, 183
Zhengda Investment Group, 75
Zhengda Research Center, 75, 80–82
Zheng Hong, 203
zhongnan qingnü (look up to men and look down on women), 178
zhongyang jingshen (central spirit), 144
Zhu Ming, 49
Zhu Xueqin, 64
ziyou (freedom), 178
Žižek, Slavoj, 216
zizhi qu (autonomous region), 91
zizhi xian (autonomous country), 91
zizhi zhou (autonomous prefecture), 91